Volcanoes

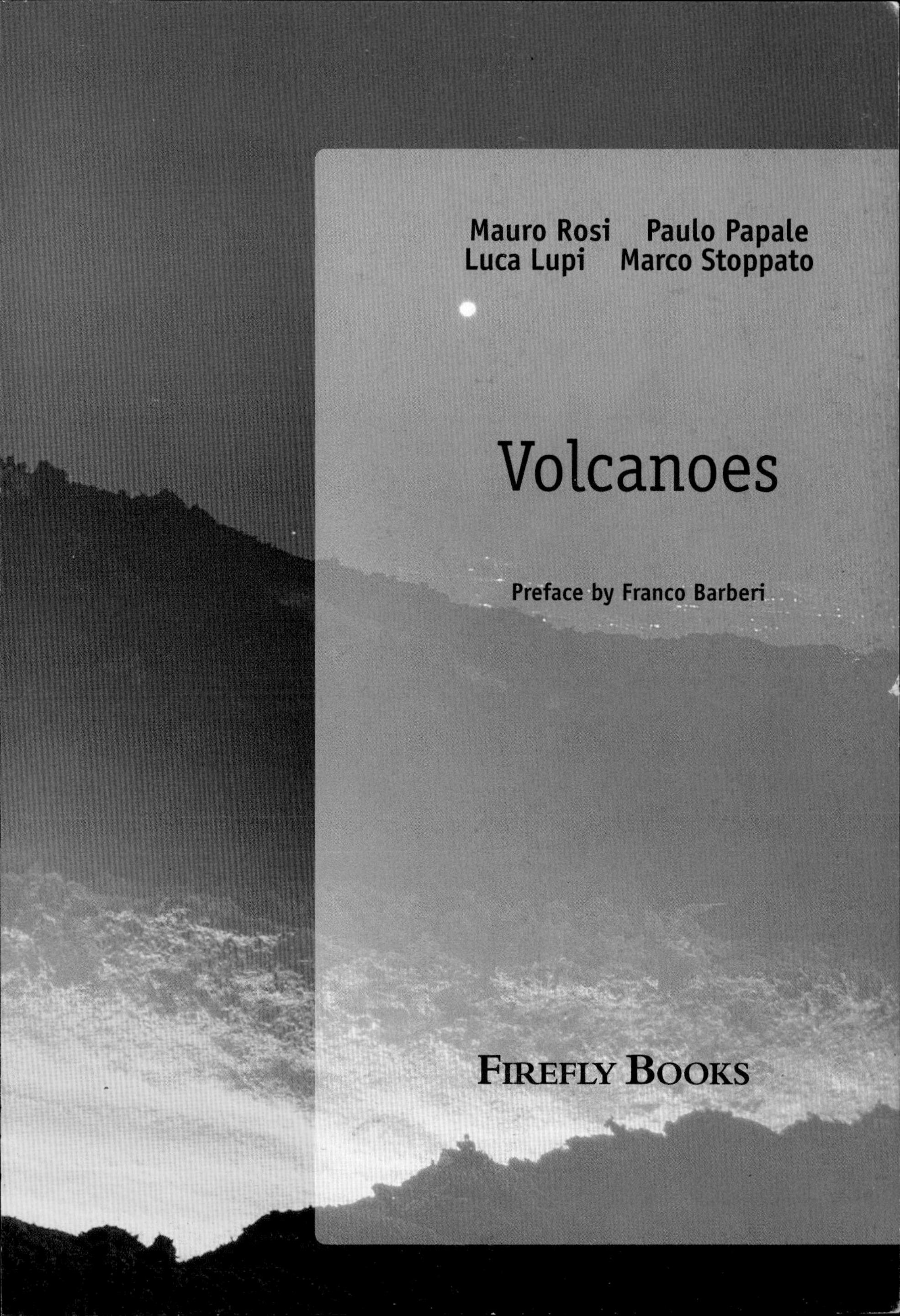

Mauro Rosi   Paulo Papale
Luca Lupi   Marco Stoppato

Volcanoes

Preface by Franco Barberi

FIREFLY BOOKS

Pages 1, 2–3, 4–5, 318–19, 320:
Mt. Etna, Sicily, 1999,
photographs by Marco Stoppato

Art director
Giorgio Seppi

Production
Mariella De Battisti

Editorial direction
Progetto Media, Milan, Italy

Layout
Barbara Capanni

Editing
Manuela Stefani

English translation
Jay Hyams

Science content advisor:
Dr. James R. Zimbelman,
National Air and Space Museum,
Smithsonian Institution.

Published in the U.S. in 2003 by
Firefly Books (U.S.) Inc.
P.O. Box 1338, Ellicott Station
Buffalo, New York 14205

Published in Canada in 2003 by
Firefly Books Ltd.
66 Leek Crescent
Richmond Hill, Ontario L4B 1H1

Printed in Spain
D.L. TO: 293-2004

A FIREFLY BOOK

Published by Firefly Books Ltd. 2003

Second Printing 2004

**Publisher Cataloging-in-
Publication Data (U.S.)**

Rosi, Mauro
 Volcanoes : a Firefly guide /
Mauro Rosi, Paolo Papale, Luca
Lupi, Marco Stoppato – 1st ed.
[336] p. : col.ill., col. photos. : cm.
Includes bibliographical references
and index.
Summary: A comprehensive
illustrated guide to 100 active
volcanoes around the world
ISBN 1-55297-683-1 (pbk.)
1. Volcanoes—Guidebooks .
I. Papale, Paolo. II. Lupi, Luca.
III. Stoppato, Marco. IV. Title
551.21 21 QE522.R67 2003

**National Library of Canada
Cataloguing in Publication Data**

Volcanoes: a Firefly guide / Mauro
Rosi . . . [et al] ; translated by Jay
Hyams.

Translation of Vulcani.
Includes bibliographical references
and index.
ISBN 1-55297-683-1
1. Volcanoes—Guidebooks.
I. Rosi, Mauro II. Hyams, Jay, 1949–

QE522.V64 2003
551.21 C2002-903937-1

CONTENTS

Volcanic eruptions, together with earthquakes, are spectacular, violent and often quite dangerous expressions of the internal dynamics of our planet.

The distribution of active volcanoes (as well as earthquakes) marks off the boundaries between and among the large rigid plates that form the surface of the Earth and that move into and away from one another. Sometimes they move away, as in ocean ridges, where frequent undersea eruptions are continuously generating new crust. At other times they converge and collide, one slipping beneath the other, forming cordilleras and island arcs. The nature of magma changes from one geodynamic environment to another and affects the type of eruptive activity, which varies from simple effusions of lava to highly dangerous explosions with clouds of gas dense with fiery fragments.

There are places on the Earth where the presence of active volcanoes has a direct effect on the social life and economy of the region. Because of their majestic beauty and the spectacular display of their eruptions, volcanoes can be enormous tourist attractions. The soil near volcanoes is often exceptionally fertile, and in many places such soil has been the reason for centuries upon centuries of flourishing agricultural activity. On the other hand, a volcano can represent a permanent threat to all settlement, to the infrastructures necessary to settled living, to human life itself. To live near a volcano requires a certain rational approach that itself requires understanding: knowing the kinds of eruption that can occur, the extent of the area at risk and the necessary measures that must be taken to prevent damage and protect life.

Many of the answers to such questions are given in this book. Written and illustrated by highly experienced experts in the field, it uses language that is simple and at the same time thoroughly scientific. It is directed at the general reader interested in knowing more about volcanoes and the Earth sciences; at the university student approaching this fascinating discipline; at the inhabitant of an area exposed to risk from volcanic eruptions. Of particular interest are the pages covering 100 of the active volcanoes of our world, located in all areas of the planet. These fully illustrated pages make this book the first comprehensive encyclopedia of volcanoes ever published. Every entry describes the principal characteristics of the volcano, such as the geodynamic environment that led to its formation, its structure, morphology, the principal types of eruption, the materials erupted, all of it based on references to historically documented events. Also provided for each volcano are directions for reaching the site and information on its most interesting aspects.

The book also addresses a theme that is currently of enormous socioeconomic and scientific importance, that of predicting eruptions and the means available to minimize the risks posed by volcanic activity. To predict an eruption one must establish the characteristics (explosiveness, area exposed to danger, sequence of events) and, well before the actual event, analyze the data collected by monitoring systems to reach a determination that an eruption is imminent. To face the threat of volcanic risk requires the advance preparation of emergency plans based on scientifically accurate scenarios of expected phenomena; also necessary is a program to educate the population of the danger. Such elements are part of the scientific and civilian measures necessary to deal with volcanic activity.

I hope this book will also find wide use as a teaching tool, most of all in grade schools and high schools, and particularly in those areas where there are active volcanoes. In such places the young should be taught early on an awareness of volcanic activity and an awareness of the danger such activity represents, but without in any way diminishing the fascination that all volcanoes possess.

Franco Barberi

INTRODUCTION

3

Hearing the word *volcano*, most people instantly imagine a cone-shaped mountain with smooth, steep slopes, perhaps with a snow-covered peak and a plume of smoke rising skyward. The description is accurate but applies to only one type of volcano, the type known scientifically as a stratovolcano. Although it is common, there are other types that differ in many ways from the stratovolcano. In place of steep slopes free of plant growth, some volcanoes are composed of long, gentle slopes, often covered with fertile soil, extending mile after mile from a central point out over the surrounding plain; such is the case with shield volcanoes. Other types have no vertical structure at all and are instead composed of a depression hundreds of yards deep extending over several miles: this type is called a caldera.

The shape of the volcanic structure is a clear indication of the type of activity carried on by the volcano itself. A highly explosive volcano throws off pumice and ash that accumulate around the crater, creating a steep-sided structure with slopes that will become deeply scored by the erosive action of water. If the volcanic activity is limited to the nonexplosive extrusion of lava—the lava flows on the surface instead of being shot into the air—the structure will have gentle slopes stretching many miles out from the source of the emissions.

A volcano's type of activity is also related to the chemical-physical properties of the magma produced, and these properties are a result of the large-scale geological characteristics of the entire region in which the volcano is located. There is in fact a deep correlation between the presence of a volcano, its type and activity, and the geodynamic structures of the Earth. To understand volcanoes, the reason why they exist in certain regions of the Earth and not in others,

Preceding pages: The eruption of Arenal in Costa Rica, June 1997. Below: The global distribution of the main tectonic plates. The borders of the plates are in red; the arrows indicate the direction of their movement. The triangles give the approximate locations of the world's active volcanoes. As can be seen, volcanoes tend to lie along the borders of plates.

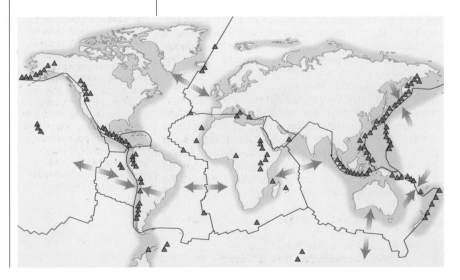

and their activity, it is necessary to study the geological forces that take place on our planet on a global scale.

THE GEODYNAMIC STRUCTURE OF THE EARTH

The surface of the Earth—its hard outer layer—has not always been the way we find it today. Over the course of geological eras, the continents have shifted their positions, creating new oceans and closing off others, pushing up towering mountain chains along the boundaries of continents, sometimes lifting ocean floors thousands of

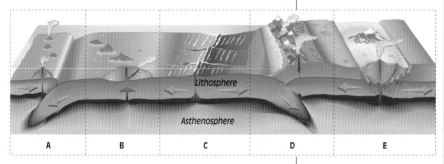

Lithosphere

Asthenosphere

A   B   C   D   E

feet above sea level. The scientific explanation of such movements, known as the theory of plate tectonics, ranks among the outstanding scientific ideas of the 20th century. According to the theory, the surface of the Earth is composed of a mosaic of separate pieces, rigid plates, which are in continuous movement in response to forces that originate inside the Earth. There are about a dozen major plates, along with smaller ones, and since each plate moves independently of the others, it happens that in some regions of the Earth they collide, or they move apart, or one slides beneath another. The majority of the planet's active volcanoes are located along such plate boundaries (see the map on the opposite page). On the basis of the geodynamic conditions that affect their morphology and activity, volcanoes can be divided into the five basic types given in the following list (the letters refer to the diagram above):

a)   island-arc volcanoes (such as those of Alaska, Japan and Indonesia);
b)   hot-spot volcanoes (such as those of Hawaii);
c)   ocean-ridge volcanoes (such as those of Iceland);

Top: Fissure at Pingvellir, Iceland, marking the edge of the American and Eurasian plates. The fissure is currently about 33 feet (10 m) wide. Above: Diagram of the geodynamic structure of the Earth. New oceanic crust is produced at the site of an oceanic ridge (C) and is subducted along a continental margin (D) or an island arc (A). The opening of a new ocean is indicated by a continental rift (E); a hot spot (B) is where a plume of magma rises to meet a moving plate. The arrows indicate the movements of individual plates.

12

d)   continental-margin volcanoes (such as the those of North America and the Andes);

e)   continental-rift volcanoes (such as those of eastern Africa).

As will be shown further on, ocean-ridge and continental-rift volcanoes are located along the boundaries of constructive plates—regions of the Earth where new crust is being created. Island-arc and continental-margin volcanoes are found along the boundaries of destructive plates, regions where one plate is being subducted beneath another to become part of the mantle. Finally, hot-spot volcanoes are located far from the boundaries of plates, which is why they are also known as intraplate volcanoes. The table below gives the yearly rates of production of magma in the constructive, destructive and intraplate volcanic zones, dividing the material into volcanic rock

ANNUAL PRODUCTION

	Volcanic rock		Intrusive rock	
	Cubic km	Cubic miles converted	Cubic km	Cubic miles converted
Environment Constructive margin	3	0.72	18	4.32
Destructive margin	0.4–0.6	0.1–0.14	2.5–8	0.6–1.92
Intraplate	0.3–0.5	0.07–0.12	1.6–3.5	0.38–0.84
Total	3.7–4.1	0.89–0.98	22.1–29.5	5.3–7.08

(rock that is produced by magma that reaches the surface in an eruption) and intrusive rock (rock that is produced by magma that has solidified within the crust of the Earth and is therefore not erupted onto the surface). As indicated by the table, most of the magma that reaches the crust of the Earth is produced by constructive plate boundaries and is not erupted onto the surface of Earth, instead being solidified in the form of plutonic rock. A total of 6 to $8^{1}/_{2}$ cubic miles (25 to 35 km^3) of magma is produced each year. This means that every hundred years a volume of new crust equal to that of a cube measuring 4 miles (6.5 km) on each side is produced by magmatic processes.

OCEAN-RIDGE VOLCANOES

The ocean bottoms are crossed by a system of deep fissures that runs for about 46,500 miles (75,000 km); basaltic magmas (see the next chapter for types of magma) are discharged through these fissures onto the ocean floor. A system of ocean ridges is located along these fissures. The fissures mark the diverging boundaries of plates, or areas where different plates are moving away from each other. Magma from the partially molten rocks of the upper mantle (the asthenosphere) rises through the fissures and is erupted onto the ocean floor and thus onto the two sides of the fissures. When it comes in contact

Opposite top: Diagram of an oceanic ridge (detail of the axial zone). The magma rises from the asthenosphere to accumulate below the ridge. Where it solidifies on the two sides it forms a layer of gabbros; where it rises it creates a series of dikes that connect the magma to the surface. Above the dikes are pillow lavas alternating with layers of sedimentary material; the farther from the axis the greater the ratio of sedimentary material to pillow lavas until, far enough away, there is only sedimentary material.

Opposite bottom: Fissure eruption at Krafia, Iceland, the surface manifestation of the Mid-Atlantic Ridge.

with water, the lava cools rapidly, forming structures called pillow lavas. Some of the rising magma does not reach the ocean bottom and solidifies within the crust, forming intrusive rocks of basaltic composition called gabbros. In this way new volcanic rock is constantly being added to the two sides of the ridge, while the oceanic plates to the right and left of the ridge expand. Over the course of time, as new volcanic rock is added, the rocks that arrived earlier are forced away from the ridge, changing as they come in contact with ocean water. Meanwhile, oceanic sediments impregnated with water accumulate above the volcanic rock, forming piles that can be tens if not hundreds of feet thick. The diagram to the right gives a cross-section of an oceanic ridge. The lava that accumulates on each side of the ridge grows to create a mountain that can at some point rise above the surface. The outstanding example of an oceanic ridge rising to the surface is Iceland, whose numerous volcanoes indicate points of activity along the ridge.

"Black smokers"

Pillow lavas

Magma

Dikes

Gabbros

CONTINENTAL-RIFT VOLCANOES

The initial stage in the opening of a new ocean is marked by the rise of the rocks that constitute the Earth's mantle. The force that causes this rise is transmitted to the rocks of the crust in the form of tensional forces that stretch the rocks and cause their collapse along the zone where the rise is taking place (see the diagram on page 14). This is the beginning of a continental rift, represented by a depression

13

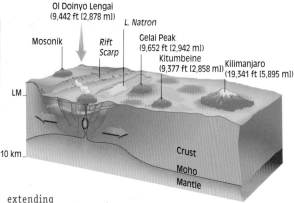

Right: Block diagram of the East Africa Continental Rift at the level of the Ol Doinyo Lengai volcano in Tanzania. The rise of the mantle causes extensional forces that arc and then break the lower crust of the Earth along a zone of fissures that run symmetrically to the axis of the ridge. The result is a depressed region in which water from the surrounding plateau will collect and along which alkaline magma rises to create a series of volcanoes.

Ol Doinyo Lengai
(9,442 ft [2,878 m])
L. Natron
Mosonik
Rift Scarp
Gelai Peak
(9,652 ft [2,942 m])
Kitumbeine
(9,377 ft [2,858 m])
Kilimanjaro
(19,341 ft [5,895 m])
LM
10 km
Crust
Moho
Mantle

extending hundreds or thousands of miles along which numerous volcanic centers can form following the rise of magma from the mantle along fissures in the crust. Such volcanic activity can be both explosive and effusive, with the presence of both stratovolcanoes and shield volcanoes. If the region of the rift is deep enough, seawater may flow into the area, leading to undersea volcanic activity. When the process reaches an advanced stage, it assumes the shape of an oceanic ridge, and the area of the rift becomes a true ocean. The most famous example of this is the Great Rift Valley in eastern Africa, the immense system of fissures that extends from the valley of the Jordan River in the Middle East across the Red Sea to Mozambique for a length of about 4,000 miles (6,500 km). A new ocean is being formed in this region, and several million years from now the region to the east of the Rift Valley will be separated from the rest of Africa, much like what happened to South America, which was part of the African continent until 130 million years ago.

Above: Crater of Ol Doinyo Lengai in Tanzania.

CONTINENTAL-MARGIN VOLCANOES

As we have seen, the volcanic rocks that form the ocean floors are progressively moved away from the ridge from which they originated. What happens when these ocean rocks, forced farther and farther from their point of origin, come into contact with the rocks that form a continent? The process that takes place is the opposite of what happens at the formation of a ridge. The dense rocks that form the ocean bottom are driven beneath the continent along what is called

a subduction zone (diagram below). As the ocean floor is reabsorbed into the mantle of the Earth, its covering of sedimentary rock and part of the underlying volcanic rocks are pushed against the continent, crumpled upward and raised, eventually forming a mountain chain. This is the way in which the great coastal chains of our world were formed; a good example is the Andes, produced by the collision between oceanic plates of the southern Pacific Ocean and the continental plate of South America.

As the volcanic rocks that constitute the oceanic floor are driven beneath the surface of the Earth, they encounter increasing temperatures. In response to this they progressively release the water contained in numerous fissures along with the water inside the minerals themselves, which have been profoundly altered through long exposure to the ocean. This water mixes with the rocks of the mantle, causing their partial melting. The end result is the production of magma that, being less dense than the surrounding rock, is driven upward by buoyancy and gas pressure to gather in reservoirs beneath the surface; this material eventually feeds the volcanoes of the continental margin. Because of the great availability of water and the particular process that produces magma, which generates liquids of high viscosity (see the following chapter), these volcanoes are characterized by prevalently explosive activity. In most cases, they are stratovolcanoes with wide calderas, such as the volcanoes

Above: Cotopaxi, Ecuador, a typical stratovolcano located on a continental margin.
Below: Diagram of an active continental-margin volcano. The oceanic crust is subducted beneath the continental. The high temperatures cause the release of water contained in the oceanic crust; this liquid mixes with the rocks of the mantle and causes them to partially melt, feeding magma to the continental-margin volcanoes. Part of the layer of sediment is pushed against the continental plate and crumbled, forming a wall of accretion separated from the ocean by a trench.

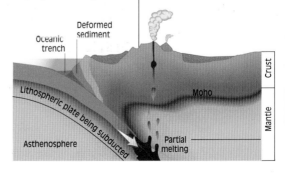

Deformed sediment

Oceanic trench

Lithospheric plate being subducted

Asthenosphere

Moho

Partial melting

Crust

Mantle

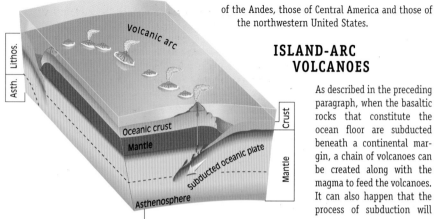

Lithos.

Asth.

Volcanic arc

Oceanic crust

Mantle

Asthenosphere

subducted oceanic plate

Crust

Mantle

Above: Diagram of an island arc. The oceanic plate is subducted under another oceanic plate. Magma rising from the mantle creates a row of volcanoes that form an island arc. Right: Explosive eruption of Sakura-Jima, Japan, an island-arc volcano.

of the Andes, those of Central America and those of the northwestern United States.

ISLAND-ARC VOLCANOES

As described in the preceding paragraph, when the basaltic rocks that constitute the ocean floor are subducted beneath a continental margin, a chain of volcanoes can be created along with the magma to feed the volcanoes. It can also happen that the process of subduction will

take place before the oceanic plate encounters a continental plate. In such situations one oceanic plate can be subducted into the mantle beneath another oceanic plate. Magma is then produced following the process described for continental-margin volcanoes except that in this case the magma will not rise through continental rocks. The absence of such rocks means that the magma can more readily reach the surface and form a volcano, which in turn means there will be a greater concentration of volcanoes. Such volcanoes come to be distributed along linear or, more often, arcuate chains, creating volcanic islands (see the diagram on the opposite page). These are island-arc volcanoes. Volcanoes of this type include the Aleutians of Alaska, the volcanic system that created the islands of Japan, the Kuril-Kamchatka islands, the Marianas-Izu islands, Papua New Guinea, the Philippines and New Zealand in the Pacific Ocean; the West Indies, between the Atlantic Ocean and the Caribbean Sea; and the Aegean arc in the Mediterranean.

Below: Diagram of a hot-spot volcano. The area where the magma rises is stationary while the oceanic plate above it is in movement. Therefore as the plate moves, its passage above the hot spot is marked by the creation of volcanoes, the line of volcanoes growing older the farther they are from the hot spot. By calculating the ages of the volcanoes, the direction and speed of the plate can be determined.

HOT-SPOT VOLCANOES

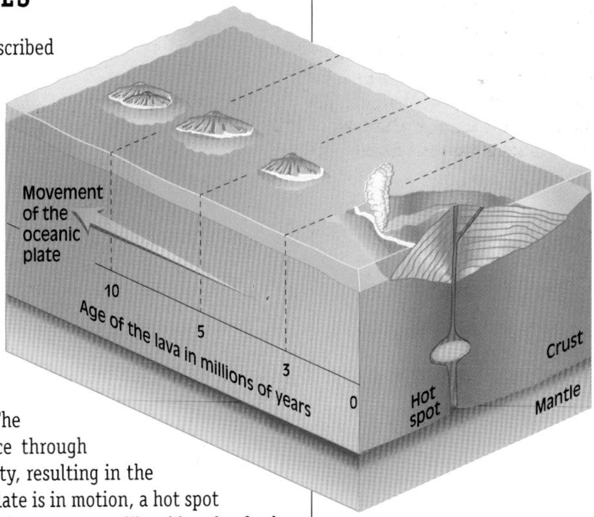

All of the types of volcanoes described so far are located along plate boundaries or in areas where a single plate has separated into two parts. There are also, however, a few regions of the Earth where magma rises through a "plume" in the mantle far from the edge of a plate. Such regions are known as hot spots. The hot spot itself is stationary, always in the same position relative to the Earth's core, but the oceanic plate above it moves. The magma that rises to the surface through the plume causes volcanic activity, resulting in the formation of islands; since the plate is in motion, a hot spot will result in a series of islands growing steadily older the farther they are from the hot spot (see diagram above).

Movement of the oceanic plate

Age of the lava in millions of years

10   5   3   0

Hot spot

Crust

Mantle

This row of islands provides an indication of the movement of the plate over the course of geological eras, also revealing any variations in the direction and velocity of the plate. This is the case in the Pacific Ocean with the alignment of the Hawaiian Islands and the contiguous Emperor seamounts and with the Galápagos Islands off the coast of Ecuador; in both cases there is a series of submerged or partially submerged shield volcanoes. Hot spots located inland (inside a continental plate) result in such volcanic activity as the hot springs and geysers in Yellowstone in North America and the volcanoes of Cameroon in Africa.

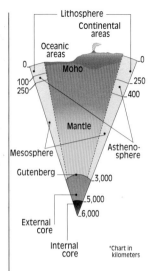

Above: The internal structure of the Earth. The dividing lines between crust, mantle and core are based on the existence of discontinuities in the speed of seismic waves. The line between the lithosphere (which includes the crust and part of the upper mantle) and the asthenosphere is based on the existence of a transition zone of low-velocity seismic waves that is believed to result from the presence of a small percentage of magmatic liquid. The convective movements that cause the movement of the plates originate within the mesosphere. The asthenosphere is the zone of plasticity on which the lithospheric plates move.

THE INTERNAL STRUCTURE OF THE EARTH

Research methods. The study of the internal structure of the Earth is based primarily on analysis of the distribution and speed of seismic waves, whether produced during earthquakes or caused by nuclear explosions. When such events occur, two types of seismic waves move through the Earth, differing in their direction of propagation. Compressional waves (also known as primary or P waves) travel in the direction of the vibration, while shear waves (also known as secondary or S waves) travel in a direction perpendicular to the vibration. While the compressional waves can move through both solids (like solid rock) and fluids (like water), shear waves cannot move through fluids. The speed at which a seismic wave moves is related to the mechanical properties, such as the relative density, of the material it moves through. Studying the arrival times of seismic waves on a global scale makes it possible to gather information on the internal structure of the Earth.

Earth's crust. On the basis of such studies, the Earth is known to have a solid upper part called the crust (diagram at left), extremely thin compared to the average radius of the Earth (3,959 miles [6,371 km]). The crust, its extent identified by an alteration in seismic waves known as the Mohorovicic discontinuity (or, more simply, Moho), has an average thickness of 19 to 25 miles (30 to 40 km) under the continents and about 4½ miles (7 km) under the oceans. It represents much less than 1 percent of the total mass of the Earth. There is a marked structural and compositional difference between oceanic crust and continental crust, a difference that reflects the different ways in which the two crusts were formed. This difference also shows up in the ages of the oldest rocks found in the two environments: in oceanic crust the oldest are never more than 240 million years old, while in continental crust they reach 3,800 million years of age. Furthermore, the structure of the oceanic crust is far simpler than that of continental crust, with rocks growing progressively older the farther they are from an oceanic ridge.

The mantle. The mantle begins below the Moho and extends nearly 1,800 miles (2,900 km) to end at another seismic alteration, this one known as the Gutenberg discontinuity, which divides the mantle from the innermost part of the Earth, the core. The mantle, which represents about 70 percent of the Earth's mass, is composed primarily of rocks in a plastic state that varies from solid to molten. The high temperature of the mantle tends to melt rocks but is contrasted by equally high pressure, which tends to keep them in a rigid state. The rocks of the mantle are liquid or solid according to the speed at which they are deformed. Under the pressure of buoyancy forces related to the different densities of different areas of the mantle, in turn produced by the increasing temperatures the closer to the inner core of the Earth, the rocks of the mantle can flow like liquids, causing slow deformation and the convective movement that drags the more superficial strata of the Earth, causing the movement of the plates. When crossed by seismic waves, which act at each point for only a brief time, the rocks of the mantle behave like a solid, permitting the passage of both type P and type S waves.

The core. The core, the innermost part of the Earth, is divided into areas separated by a further seismic discontinuity, the Lehman discontinuity. Since S-type waves cannot move through the area of the outer core above that discontinuity, it is believed to be composed of matter in a liquid state. The inner core is solid.

Asthenosphere and lithosphere. The outermost part of the Earth, composed of the crust and part of the mantle, is rigid, meaning it has the characteristics of a solid and is marked by the high speeds of seismic waves typical of solids. It is called the lithosphere and is commonly taken to coincide with the area of moving continental plates. Beneath the lithosphere is an area of the mantle, extending about 60 miles (100 km), where the speed of seismic waves decreases rapidly, only to then increase again. This region of low velocity (known as the Low Velocity Zone, LVZ) is the asthenosphere. In this area the conditions of pressure and temperature are such to cause the partial melting of the rocks of the mantle and thus the production of magmatic liquid. The lower portion of the mantle under the asthenosphere is called the mesosphere.

Below: Table giving the average composition of the crust, mantle and core. The values indicate the percentages by weight of each component (from Gasparini and Mantovani, Fisica della Terra Solida, Naples: Liguori Editore, 1984; page 509).

19

Oxides	Continental crust	Oceanic crust	Mantle	Core	
SiO_2	60.2	48.4	43.4	Fe	84
TiO_2	0.7	1.3		Si	11
Al_2O_3	15.2	15.9	3.9	Ni	5.3
Fe_2O_3	2.5	2.7			
FeO	3.8	6.1	9.3		
MnO	0.1	0.2			
MgO	3.1	6.9	38.1		
CaO	5.5	12.2	3.7		
Na_2O	3.0	2.6	1.8		
K_2O	2.9	0.5			
P_2O_5	0.2	0.1			
C	0.2				
CO_2	1.2	1.9			
S	0.1				
H_2O	1.4	1.3			

MAGMA

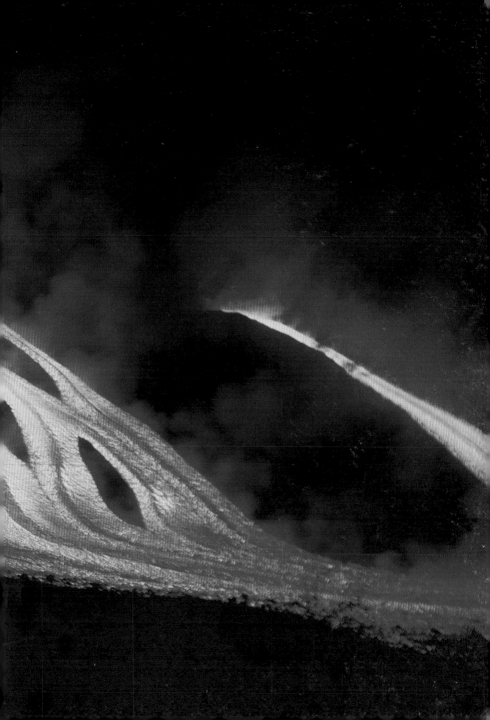

Magma is molten or partially molten rock below the surface of the Earth; when it erupts onto the surface it is called lava. Magma is made up of numerous components and usually goes through several states, being a liquid, solid or gas. Several variables contribute to determining the type of magma erupted from any given volcano; there are differences in the magma erupted from a volcano during the course of different eruptions, even differences in the magma erupted during different phases of a single eruption. The most important of these variables are the magma's composition and temperature, its content in crystals or in solid fragments of various nature, and the amount of gas. These variables lead to different properties of magma in terms of density, viscosity and ability to contain dissolved gas. These variables have an enormous effect on the type of eruption, on the characteristics of the volcanic deposits and finally on the structure of the volcano itself.

CHEMICAL COMPOSITION

From the chemical point of view magma is an extremely complex system. The chemical composition of magma does not remain constant over time but varies in response to variations in the environment in which it is located. In contact with colder rocks, magma loses heat. This change in temperature causes certain

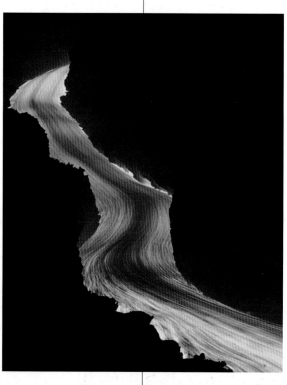

Preceding pages: Rivers of lava flow down the slopes of Mount Etna, Sicily.
Above: A lava flow on Etna.
The darker areas in the middle are places where the lava is beginning to form a crust as a result of cooling.

minerals to begin to crystallize, depriving the magma of those chemical components that are involved in the formation of the crystals while at the same time enriching the magma with other components that are not involved in the formation of the crystals. Aside from heat, magma also exchanges chemical components with the surrounding rock, modifying both the composition of the magma and the surrounding rocks. Portions of rock can be incorporated into the magma, becoming molten or remaining as solid fragments within it. In response to variations in chemical composition, temperature and most of all pressure, volatile substances contained in the magma like water or carbon dioxide—present in magma in significant quantities and initially dissolved in it—can be released to form gas bubbles, producing great changes in the properties of the magma and in many cases leading to an eruption.

MAGMATIC SERIES

From what has been said it should be clear that magma can be looked upon as a system in continuous evolution.

The geodynamic environment in which magma is produced strongly affects its characteristics. To a large extent, the original environment of magma gives it a sort of initial chemical "signature" that will prove determinant through the magma's later stages of crystallization and evolution. For these reasons, study of the chemical and mineralogical composition of a volcano's products allows scientists to assemble a great deal of information concerning the evolutionary history of its magma.

This fact makes possible the definition of a magmatic series, meaning a sequence of magmas of differing composition but belonging to the same geodynamic environment and therefore characterized by the same chemical "signature," so that, ideally, the series can be considered different stages in the evolution of the same magma.

Based on the various geodynamic environments described in the preceding chapter, the principal magmatic series are:

a) tholeiitic series, typical of an ocean ridge;
b) alkali series, typical of a continental rift or hot spot;
c) calc-alkaline series, typical of a continental margin or island arc;
d) potassic series, typical of a convergent margin in an advanced stage of evolution.

In addition to these series there are also the magmas produced through the partial melting not of rocks of the mantle but of rocks of the crust, at a depth at which the conditions of pressure and temperature favor the production of significant quantities of liquid. Such magma is known as anatectic magma; it is found in many places in Italy, for example, on the island of Elba (the granite of Monte Capanne) and in southern

Above: Flow of carbonatite lava at Ol Doinyo Lengai, Tanzania. The lava is flowing from the small cone, or hornito, in the background; it is dark when it appears but rapidly turns white following contact with the atmosphere.

Below: Flow of carbonatite lava at Ol Doinyo Lengai. The composition and temperature of carbonatite lava differs from those of silicate lavas. It is extremely fluid and dark in color.

Below: Examples of three types of magmatic rock; each piece is about 4 inches (10 cm) long. Beneath each sample is an image of the same rock shown in thin section using a polarizing optical microscope (crossed polarizing prism, horizontal size between ⅖ and ¾ inches [1 and 2 cm]). From left to right: Granite, gabbro, basaltic lava. Note the difference in texture between the granite and the gabbro, which are intrusive rocks composed entirely of crystals, and the lava, which is a volcanic rock composed of crystals immersed in a microcrystalline bed.

Tuscany (the rhyolites of San Vincenzo in the province of Leghorn).

A further category is the carbonatite magmas, which differ from all the others. The primary component of the magmatic series described above is silica (SiO_2), for which reason they are called silicate magmas; whereas calcium carbonate ($CaCO_3$) is the principal component of the carbonatite magmas, which are produced by processes of chemical mixing beginning with magma rich in carbon dioxide.

Furthermore, while the temperature of silicate magmas varies from 1472°F to 2192°F (800°C to 1200°C) according to its composition, the temperature of carbonatite magmas is much lower, usually ranging between 932°F and 1112°F (500°C and 600°C). There were carbonatite volcanoes in Italy in the distant past, in the regions of Umbria and Latium. At the present time the only active carbonatite volcano in the world is Ol Doinyo Lengai in Tanzania (see Number 22, pages 152–53).

The chemical-physical conditions in the area where a certain magma is produced determine the type of magmatic series. The rocks of the mantle that are the source of most of the magma that erupts onto the surface are characterized by a substantial homogeneity of chemical composition (see the table on page 19). Despite this initial homogeneity, however, the liquids that are produced through the partial melting of the rocks differ according to the environment in which the melting takes place, whether in an ocean ridge, a continental margin or other environment.

In an ocean ridge, where the mantle rises almost to the surface, a good portion of it (about 20 percent) lies in a low-pressure zone. Beneath a continental rift, on the other hand, and even more so at a hot spot, the partial fusion of the mantle takes place at a far greater depth, with far greater pressure, and involves only a small part (less than 5 percent) of the original rock.

Finally, in the situation presented along the boundary of convergent plates, whether continental or oceanic (island arc), the

partial melting of the mantle takes place principally as a result of water released from the oceanic plate, which is in turn incorporated into the mantle itself.

INTRUSIVE ROCKS AND VOLCANIC ROCKS

In the Introduction, mention was made of the difference between intrusive and volcanic magmatic rocks. In fact, only a small portion of

the magma that rises through the crust is thrown onto the surface through volcanic activity. Most of the magma remains trapped within the crust, where it slowly cools, forming intrusive magmatic rocks. The slow cooling permits the complete crystallization of the intrusive rocks, the crystals of which are therefore large, often easily visible to the naked eye. Such is the case with granite in continental settings and of gabbros in oceanic settings. Such large crystals do not form, however, if instead of

Above: Lava structures at the Baia dei Mostri on the island of Vulcano in the Aeolian Islands.
Below left: Pumice deposit produced by explosive activity and the same material viewed under a polarizing microscope (parallel polarizing prism, horizontal size around ⅖ of an inch [1 cm]). Pumice is composed of numerous cavities (vesicles) immersed in a matrix composed primarily of volcanic glass.

getting trapped in the crust the magma rises and is thrown onto the surface, for part of it is still liquid when it reaches the surface. In contact with the atmosphere, it cools rapidly, and this rapid cooling does not permit the formation of large crystals throughout the rock. Such magma becomes volcanic rock comprised of much smaller crystals (crystals that are microns in size); or in many cases the magma liquid congeals so rapidly that crystals have no time to form, and it hardens into a kind of glass (obsidian). This is the case of the lavas and pumice given off during explosive-type eruptions. The larger-size crystals, which were formed within the crust before the eruption, are in most cases visible to the naked eye and are called phenocrysts.

THE CHEMICAL EVOLUTION OF MAGMA

From the moment of its production to the moment of its eruption, or of its solidification within the crust of the Earth, magma undergoes important chemical and mineralogical changes. The longer the time that passes between the magma's production and its eruption or solidification, the greater the extent of such changes. Many factors are involved in this process, but the one that is thought to be the most important is the process of fractional crystallization. This process consists of the precipitation of crystals and their separation from the magmatic liquid. Since each kind of crystal, as it forms, selectively impoverishes the magmatic liquid of those elements that go to form the crystal, the result is a gradual variation in the composition of the magma as new crystals are precipitated and separate away from it.

The process that leads to the separation of the crystals from the liquid magma is related to their different densities and to the modalities of cooling and crystallization of the magmatic bodies in the crust. Crystals that are denser than the surrounding liquid tend to sink in the magma and accumulate in the deeper areas of the magmatic body. Crystals that are less dense than the liquid may rise upward under the force of buoyancy and form accumulations in the upper part of the magmatic body. Contact with surrounding rocks diminishes the temperature of the magma, increasing the speed of the crystallization, which in turn effects the chemical evolution of the magma. The table on page 19 shows the average composition of the rocks of the mantle. Aside from those given in the table, other components are present in trace amounts. The processes of partial melting and fractional crystallization affect the amounts of those elements that will remain in the magma to become important constituents of the magma. For example, potassium is extremely scarce in the rocks of the mantle, but it can become abundant in magma from the mantle, such as those belonging to an alkaline magmatic series. The process of chemical contamination with the rocks of the crust can also make a substantial contribution to enriching the magma in terms of elements that were initially present only in small quantities.

Above: A small flow of basaltic lava from Kilauea in the Hawaiian Islands.

CRITERIA OF CLASSIFICATION OF MAGMATIC ROCKS

One of the simplest means of evaluating the stage of chemical evolution of magma is measurement of its silica (SiO_2) content. With the exception of certain extremely rare types, such as carbonatite, silica is the most abundant element in magma, and the process of fractional crystallization tends to produce a progressive enrichment of that component. On the basis of the silica content, it is possible to distinguish four categories of magma:

Ultrabasic magma	$SiO_2 < 45$ percent by weight
Basic magma	$45 \leq SiO_2 < 52$ percent by weight
Intermediate magma	$52 \leq SiO_2 < 62$ percent by weight
Acidic magma	$SiO_2 \geq 62$ percent by weight

This division is applied both to magma and to the rocks resulting from the solidification of magma. In addition to this, the criteria used for the further classification of rocks (intrusive or volcanic) are different. In the case of intrusive rocks, complete crystallization permits a division based on type and abundance of each kind of mineral present. In the case of volcanic rocks, the more or less abundant presence of a mass of smaller crystals or of glass, or the incomplete or completely absent crystallization of the magma, means that classification on the basis of the minerals is not applicable. Consequently, the classification of such volcanic rocks is carried out on the basis of the chemical composition of the rock.

The various criteria used in the classification were established by the International Union of Geological Sciences in 1973 (IUGS Subcommission on the Systematics of Igneous Rocks, Classification and Nomenclature of Plutonic Rocks, 1973).

CLASSIFICATION OF MAGMATIC INTRUSIVE ROCKS

The diagram from the IUGS convention of 1973 reproduced here is based on the relative amounts of minerals grouped in five different symbols:

Q	=	quartz
A	=	alkali feldspar (albite, orthoclase)
P	=	plagioclase (except albite)
F	=	feldspathoids
M	=	mafic minerals (olivine, pyroxene, amphibole, biotite and accessories)

Left: Table giving the silica (SiO_2) content of four types of magma, from ultrabasic to acidic.
Below: Strachaisen classification diagram (simplified) for intrusive magmatic rocks. The letters Q, A, P and F at the four points indicate the relative abundance (by volume) of the minerals or classes of minerals quartz, alkali feldspar, plagioclase and feldspathoids. The rocks that fall in the upper area of the diagram are saturated with silica (SiO_2), while those that fall in the lower part are scarce in silica.

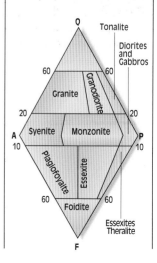

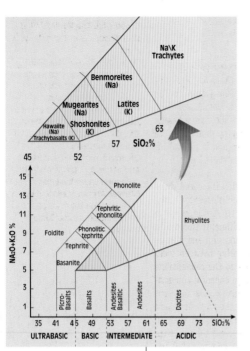

A description of the various minerals in magma goes beyond the scope of this book. To understand the criteria used in classification it is sufficient to bear in mind the following points:

1. The minerals corresponding to the symbols Q, A, P and F (all except the mafic) are called sialic. They are the primary components of acidic intrusive rocks and constitute the great majority of intrusive continental rocks.

2. The mafic minerals (M) do not appear in the classification diagram, but when present in significant quantities they are used as a basis of classification, adding a second name to the classified rock on the base of the sialic minerals (for example, a granite can be called a pyroxene or an amphibole according to the dominant mafic mineral).

3. Quartz (Q) and the feldspathoids (F) are incompatible: In fact, the feldspathoids form only if silica is present in scarce quantities, while quartz (pure silica) needs large quantities of silica to crystallize.

4. The alkali feldspars need large quantities of sodium and potassium (alkaline elements) and are therefore typical of the alkaline magmatic series, which are only scarcely found among intrusive magmatic rocks.

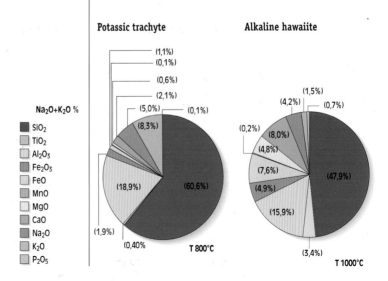

Potassic trachyte

Alkaline hawaiite

Na$_2$O+K$_2$O %

- SiO$_2$
- TiO$_2$
- Al$_2$O$_3$
- Fe$_2$O$_3$
- FeO
- MnO
- MgO
- CaO
- Na$_2$O
- K$_2$O
- P$_2$O$_5$

T 800°C

T 1000°C

CLASSIFICATION OF VOLCANIC MAGMATIC ROCKS

The diagram from the IUGS, reproduced on the opposite page, is based on the abundance of silica (SiO_2) and the alkalies (Na_2O and K_2O). For this reason it is known as the TAS (Total Alkali-Silica) diagram. The calc-alkaline magmatic series common to convergent-margin situations is located beneath the lowest red line and includes rocks from basalts (scarce) to andesites, dacites and rhyolites (extremely abundant). The alkali series is located between the two red lines and is marked off according to the most abundant alkali element in the alkaline-sodic and alkaline-potassic series. The potassic series common to a convergent margin in a highly advanced stage of evolution is located above the highest red line and includes tephritic to phonolitic rocks. Finally, in the tholeiitic series common to oceanic ridges, far and away the most common types of rocks are basalts. Further differentiation can be made on the basis of other chemical elements, and iron and magnesium are of particular importance for discriminating between rocks with tholeiitic affinities and those more similar to the calc-alkaline series. The five pie charts below give the average chemical composition of magmas from several geodynamic environments and different stages of evolution. As indicated by the charts, the most abundant element of magmas is silica (SiO_2), which varies from less than 50 percent by weight in basalts (chemically less evolved) to more than 70 percent in rhyolites (chemically evolved). Aluminum (AL_2O_3) is most often the second most abundant element, but in weakly evolved magmas, such as picrites, it can be exceeded by magnesium oxide (MgO). The oxides of iron, calcium, sodium and potassium are other

Opposite top: The TAS (Total Alkali-Silica) Classification used for volcanic rocks.
Below: The chemical compositions of five magmas from different geodynamic settings and at different stages of evolution. Based by the values given in percent we see that the principal component is silica (SiO_2), which varies from about 50 percent in basalts to more than 70 percent in rhyolites. Ultrabasic magmas, such a picrites, which do not appear on the table, contain between 40 percent and 50 percent silica (from D'Amico et al., Magmatismo e metamorfismo, Turin: UTET, 1987).

Calc-alkaline rhyolite

(0,7%)
(0,1%)
(0,8%)
(0,9%)
(2,1%)
(4,2%)
(3,3%)
(0,1%)
(14,1%)
(72,5%)
(0,2%)

T 750°C

Calc-alkaline andesite

(0,1%)
(3,2%)
(3,4%)
(1,3%)
(0,2%)
(7,0%)
(17,3%)
(59,2%)
(4,5%)
(3,0%)
(0,8%)

T 1000°C

Tholeiitic basalt

(0,2%)
(2,7%)
(0,1%)
(11,4%)
(7,7%)
(9,9%)
(50,7%)
(15,6%)
(1,5%)

T 1100°C

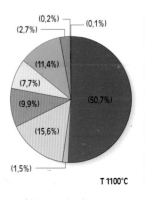

components that can be present in important quantities. Their relative abundance depends on the type of volcanic series and the state of evolution of the magma so they make distinctions possible among differing magmas in different geodynamic environments.

Above: Gas emanation from the summit crater of Etna, Sicily. Below: The average composition of volcanic gases. The most abundant component is water, followed by carbon dioxide and sulfur compounds.

MAGMATIC GASES

Aside from the elements given on the preceding pages, magmas contain numerous kinds of volatile substances, components that, according to the conditions, can be dissolved in liquid magma or be separated from the liquid to form a gas. Although often present in relatively small quantities, such elements are extremely important because in many cases it is the presence of a gas phase that will cause a volcanic eruption and determine its dynamics. During an explosive eruption, from 10 million to 1,000 million tons of volcanic gas are discharged into the atmosphere over a period of time running from a few hours to a few days. A similar quantity of gas is produced annually through the fumarolic activity of volcanoes like Sicily's Mount Etna and is the result of the slow release of gas from magma inside

H_2O

Altro

CO_2

H_2

SO_2

HCl

H_2S

Tracce

the volcano. Such quantities of volcanic gas have an effect on the atmosphere on a worldwide scale, contributing to changes in its composition and in many cases producing unwanted effects like acid rain. The table below gives the type and relative amounts of volatile compounds in two types of basaltic magma, tholeiitic and alkaline. As can be seen, the composition of magmatic gas depends on the type of magmatic series, reflecting the role of the processes of the genesis of magma in different geodynamic environments.

Water is the most abundant type of volatile compound; it varies from hardly more than 1 part to 7–8 parts by weight percent, according to the magma's stage of chemical evolution. In fact, the processes of differentiation that cause the chemical evolution of a magma imply, in general, an increase in the concentration of the water and of the other volatiles, since these are only scarcely incorporated in the miner-

	CO	CO_2	HCl	H_2	H_2O	H_2S	SO_2	S_2
Tholeiitic basalt	0.39	5.5	0.43	2.9	87.4	0.54	2.7	0.09
Alkaline basalt	0.51	24.3		0.53	49.3	0.20	24.9	0.21

als. In terms of its abundance, the second most important volatile compound is carbon dioxide (CO_2), followed closely by sulfur dioxide (SO_2).

Other kinds of volatile compounds present in appreciable quantities are carbon monoxide (CO), hydrochloric acid (HCl), hydrogen (H_2), hydrogen sulfide (H_2S) and sulfur (S).

PROPERTIES OF MAGMAS

The large chemical variability of magmas and the extremely variable conditions of pressure and temperature in which they exist, from the upper mantle to the surface of the Earth, are reflected in the broad spectrum of properties found in magma, such as density, viscosity, thermal conductivity and capacity to hold in solution such volatile compounds as water or carbon dioxide. Such properties are of enormous importance in volcanic processes and have a powerful influence not only on the type of eruption and the type of material produced,

Top: Fumarole and incrustations of sublimates of sulfur at Vulcano (Aeolian Islands, Italy).
Above: Table giving the composition of magmatic gases for two types of basalts: tholeiitic (Surtsey, Iceland) and alkaline (Etna, Sicily). The values are given in molar percent, units of measure that indicate the abundance on the molecular base of gas (from J.R. Holloway, "Volatile Interactions in Magmas," in R.C. Newton, Thermodynamics of Minerals and Melts, New York: Springer Verlag, 1981).

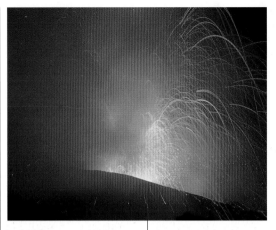

Above: Strombolian eruption, a type of eruption associated with magmas of medium-low viscosity. Below: Flow of block lava in Iceland. The outer surface of a lava flow can vary a great deal and is related to the chemical properties of the lava itself.

but on the likelihood that an eruption will occur at any given moment.

SATURATION BY VOLATILE ELEMENTS

The formation of gas bubbles in a magmatic body, which remains for a certain period inside the crust, is a mechanism able to trigger a series of processes that, in many cases, ends in an eruption. That happens because the formation of gas bubbles is related to variations in the pressure of the magma, which can thus open the way toward the surface, fracturing the rocks above. Quite clearly, then, the chance of an eruption is directly related to the ability of a given body of magma to maintain gas in solution, avoiding the formation of gas bubbles.

The level of pressure and temperature at which magma begins to form gas bubbles in chemical balance with the liquid from which the bubbles escape is called the state of saturation. This depends on the chemical composition of the magmatic liquid and on the type and abundance of the dissolved volatile compound. Of those factors, pressure is the basic parameter that determines how much gas is released and how much goes to form gas bubbles. As pressure diminishes, increasingly small quantities of water, carbon dioxide and other magmatic gas can remain dissolved in the liquid. As a result, more and more gas bubbles are formed, or the already existing bubbles increase in size. Since the pressure diminishes as the magma moves from inner areas of the Earth toward the surface,

the magma will be increasingly rich in bubbles as it rises along a volcanic conduit. This is a process of enormous importance in volcanic eruptions and is responsible for numerous processes that will be dealt with in greater detail in the next chapter.

DENSITY OF MAGMA

Above: A large gas bubble explodes on the surface of extremely fluid Hawaiian lava.

The density of a body, meaning the relationship between the mass of the body and the volume that it occupies, is a measurement of extreme importance. A body immersed in a fluid that is denser than itself tends to float; if the fluid is less dense, the body will tend to sink. Since magmas are bodies that go through different phases, from solid to liquid to gas, it is clear that the relative density of these various phases will determine where they will concentrate, whether toward upper or lower areas.

Gas bubbles, for example, being lighter in weight than the magmatic liquid that surrounds them, tend to move upward through the magma. This process explains why the surface of a body of lava often seems bubbly, and it is also the reason why the gas initially contained in a magmatic body inside the Earth will free itself from the magma and reach the surface, forming fumaroles. As for the minerals in the magma, there are some that are denser than the magma and others that are less dense. In general, minerals rich in iron and magnesium, such as olivine, pyroxene and opaque oxides

Below: The basalt composition and high temperature of the magma emitted by Kilauea, Hawaii, are the cause of its low viscosity, which in turns causes lava fountaining such as this.

like magnetite and ilmenite, are denser than magmatic liquid and therefore tend to sink.

Since these minerals are formed in abundance during the first phases of the chemical evolution of magma, their sinking—and thus their separation from the liquid portion of the magma—impoverishes the liquid of iron and magnesium and contributes to the chemical evolution of the magma by means of fractional crystallization (see page 26). On the other hand, minerals such as the plagioclases (rich in aluminum, sodium and calcium) and the feldspathoids (rich in aluminum, sodium and potassium) are often less dense than the surrounding liquid and will rise upward (the process of buoyancy) creating accumulations at the top of the magmatic body.

Below: The basalt composition and high temperature of the magma emitted by Kilauea, Hawaii, are the cause of its low viscosity, which in turns causes lava fountaining such as this.

VISCOSITY OF MAGMA

More than any other property, it is probably the viscosity of magma that determines its behavior during an eruption. Viscosity represents the measure of the capacity of a fluid to deform and therefore to flow, and its importance is based on the fact that magmas can present viscosities that differ by hundreds or thousands of millions of times according to the conditions in which they are found. Since the dynamics of an eruption depend to a large measure on the way the

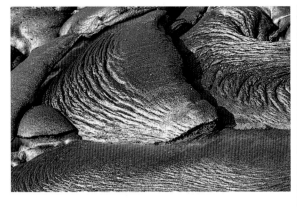

Left: Very fluid lava results in smooth-surfaced lava flows known as pahoehoe.
Below: The molecular structure of a magmatic liquid affects its level of polymerization, which is an important factor in its level of viscosity.

magma rises along the fissures and conduits that channel it from the depths of the Earth to the volcanic vent, it follows that the relative viscosity of the magma will determine the kind of eruption. In effect, there is a general relationship between viscosity and explosivity in volcanic eruptions, just as there is one between viscosity and the dynamics of a lava flow. These subjects are considered in greater detail in the next chapter. The viscosity of a magma depends on numerous factors, including its chemical composition, temperature, amount of dissolved water and amount of crystals and gas bubbles.

THE ROLE OF THE CHEMICAL COMPOSITION OF MAGMATIC LIQUID

As has been said, a magmatic liquid is a multi-component system whose various elements, in appropriate conditions of pressure and temperature, form crystals of various types. However, even at the liquid state magma contains rough blueprints for the complex structures that will later appear in crystals. In other words, various molecules or groups of molecules join to form structures such as monomers, dimers, chains or polymers (see the figure at right) that restrict the opportunities of the molecules to move freely. Since a reduction in the opportunities for molecular movement will diminish a liquid's ability to deform in response to the action of any given force, the presence of such structures is reflected in increased viscosity. The various oxides that compose magmatic liquids perform differing roles in the construction of molecular

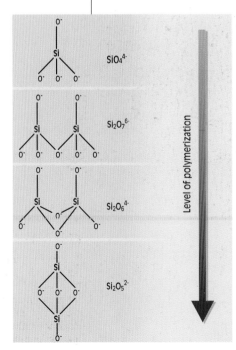

SiO_4^{4-}

$Si_2O_7^{6-}$

$Si_2O_6^{4-}$

$Si_2O_5^{2-}$

Level of polymerization

Below: The viscosity of magma varies by many orders of magnitude according to its composition and temperature.

structures, or in the ability of a liquid to polymerize. Some oxides contribute to the formation of molecular chains, others interrupt them. Silicon is the most efficient element at forming structures; the higher the silicon content, the greater the polymerization of a liquid and thus the greater its viscosity. Aluminum is also a builder of structures, except at high pressure, when its behavior changes and it diminishes the ability of a magmatic liquid to polymerize. Elements like iron and magnesium modify structures; they diminish the ability of polymerization and therefore the viscosity of the magma. The effect of water on viscosity depends on its state. When present as a dissolved liquid in magma it alters the efficacious creation of structures by diminishing viscosity. When it passes from a liquid to a gas near the surface, however, it increases viscosity.

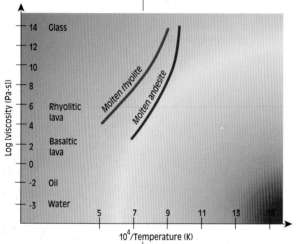

THE ROLE OF TEMPERATURE

The temperature of a body is an average measure of the body's internal energy. An increase in temperature corresponds to an increase in internal energy, and this increase happens by way of an increase in the amount of vibrations of each atom. Since the more an atom vibrates the more mobile it is, an increase in temperature leads to a reduction in viscosity. This reduction is extremely efficient, as shown in the diagram to the left, since it involves the entire structure of the liquid. For example, an increase in temperature from 2012°F to 2282°F (1100°C to 1250°C) corresponds to a decrease in the viscosity of more than a thousand times for a magma of andesite composition and more than ten thousand times for a magma with a rhyolite composition.

THE ROLE OF THE PRESENCE OF CRYSTALS

The presence of crystals represents an obstacle to the flow of liquid magma, since the strata of liquid in movement are forced to adapt to the presence of the crystals by way of further deformation, which increases the friction of each stratum. As a result, the presence of crystals constitutes a further increase in the viscosity of magma. The more crystals, the greater the increase in viscosity; for example, the

Left: Lava fountain with low-viscosity lava on Hawaii.
Below: Bomb of basaltic lava, its interior still incandescent, on Stromboli, Italy.
Bottom: Lava pouring into the ocean on Kilauea, Hawaii, causing large quantities of steam.

viscosity of a magma with 35 percent of crystals by volume is about 10 times greater than that of the same magma at the same temperature without crystals.

THE ROLE OF THE PRESENCE OF GAS BUBBLES

Gas bubbles in magma have the identical effect as crystals, except for the fact that in certain conditions they can change shape. This peculiarity constitutes a great difference; in fact, in cases in which the gas bubbles do not change shape by the flow of the magma, they contribute to increasing its viscosity. On the contrary, in cases in which the gas bubbles change shape, they create areas where the friction among the various levels of the magma is more or less nullified because of the very low viscosity of the gas compared to that of the magmatic liquid. In such cases, the effect of the presence of gas bubbles is that of reducing the viscosity of the magma. In general, other conditions being the same, the greater the viscosity of a magmatic liquid, the more likely it is that gas bubbles will change shape, and the greater the radius of each bubble. In numerous cases, it is therefore possible that small gas bubbles will not change shape, producing increased viscosity, while larger bubbles will change shape, producing a decrease in the viscosity of the magma.

VOLCANIC ERUPTIONS

Without a doubt, volcanic eruptions are among the most spectacular natural phenomena to occur on our planet. At the same time, however, they are often extremely dangerous events to humans. Of course, volcanic eruptions vary greatly in the amount of material erupted and its effect on the surrounding environment. Some eruptions involve the relatively peaceful emission of small amounts of material with only slight effects on the environment. Other eruptions are true cataclysms that discharge enough ash and gas into the atmosphere to disturb the climate, devastate the flora and fauna of entire regions and lead to sudden, radical changes in the environment.

Volume of Material Emitted

50 km^3 — Tambora (Indonesia) 1815
Santorini (Greece) 1500 B.C.

20 km^3 — Krakatau (Indonesia) 1883

15 km^3 — Katmai (Alaska) 1912

4 km^3 — Vesuvius (Italy) 79 A.D.
Pinatubo (Philippines) 1991

1 km^3 — St.Helens (USA) 1980

0.5 km^3 — Vesuvio (Italy) 1631
Mount Pelée (Martinique) 1902
Mount Etna (Italy) 1991–93

As explained in the Introduction, volcanic eruptions take place both on the surface of the planet and at the bottom of the oceans. Underwater volcanic activity emits from three to five times as much material as does surface volcanism (see the table on page 12). This is only logical since there are roughly 1,000 volcanic islands in the Pacific Ocean, but the number of volcanoes on its bottom is estimated to be on the order of 55,000. Statistics indicate that the number of continental and island volcanoes that produce eruptions annually is between 50 and 60.

To understand the factors that lead to a volcanic eruption it is first necessary to identify and define the elements that comprise a volcanic system. A volcano is composed of three basic elements: the vent through which volcanic materials erupt; the conduit (the passage along which magma reaches the surface) and the magma chamber, or reservoir, where magma is stored. The magma chamber is the volcano's primary engine; it is located inside the crust of the Earth, beneath the structure of the volcano itself. Here the magma is concentrated and eventually transformed before being erupted.

The depth, shape and size of the magma chamber vary a great deal from volcano to volcano. In general, the magma chamber is formed inside the Earth's crust at a depth of 2 to 2½ miles (3 to 4 km) or more from the surface. There have been only a few rare cases of magma chambers located inside a volcanic structure itself, or at a high level within it.

The conduit is the canal through which the magma rises. It is believed that at great depths the conduit is a fracture in the Earth and would therefore have an elongated section; but at upper levels the conduit is thought to take a more cylindrical shape, flaring toward its upward end, or throat.

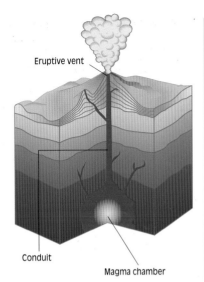

Above: The principal elements of a volcanic system are the magma chamber, the conduit and the eruptive vent.

The vent is the opening from which volcanic materials are discharged. An eruption takes place when magma, stored in the chamber, reaches sufficient pressure to overcome the resistance of the rocks that form the roof of the magma chamber. In most cases an eruption will bring to the surface only a portion of the volume of the magma stored in the magma chamber. This magma can be discharged by way of effusive or explosive volcanic activity. Effusive activity consists in the overflowing of degassed magma from a crater or vent of the volcano with the consequent formation of a lava flow. Explosive activity can lead to the complete destruction of the magma chamber and the discharge into the atmosphere of fragments of magma at high temperature. Many eruptions are exclusively effusive or explosive, but an equal number involve both explosive activity and, at the same time, effusive flows of lava.

EFFUSIVE ACTIVITY

The arrival on the surface of magma that has lost its gas component produces the formation of a lava flow or of an accumulation of lava around the vent.

The movement of lava is regulated by gravity. The relative steepness of the slope with distance and the speed at which the lava cools are among the factors that slow the flow and eventually cause it to stop. The speed at which lava flows and the distance covered are affected by the viscosity of the lava, the angle of the slope and the amount of material involved in the eruption, i.e., the quantity of magma emitted over a certain period of time.

Low-viscosity basalt lava can be fast moving, particularly when abundantly fed and running down an inclined plane. Such low-viscosity lava moves quickly and can cover tens and even hundreds of miles.

To give a single example, the 1989 lava flow from Mount Etna, Sicily, moved at a velocity between 52 feet (16 m) a second the first day to 1 foot (0.3 m) a second in the final phase; lava flows from Hawaiian volcanoes tend to have a lower viscosity than those from Etna and usually move faster. In 1855, the flow from Mauna Loa reached speeds of 40 miles (64 km) per hour on a slope with an average angle of 10–25 degrees.

If highly viscous magma is erupted at a slow enough rate it may pile up above the vent rather than flow away from it, leading to the formation of steep-sided mounds of lava known as lava domes. Such effusive activity can go on for many years or even decades. For example, since the 1956 paroxysmal eruption of the Bezymianny volcano in Kamchatka, lava has continued to flow from the vent, forming a dome. Many lava domes grow to large sizes, their steep sides covered with rock debris.

PRODUCTS OF EFFUSIVE ACTIVITY

Effusive activity leads to the formation of various types of lava material. In general, lava has a compact appearance, although its upper surface may be covered with bubbles, with chunks of scoria, called clinkers, or with a chaotic mass of lava blocks. Based on its chemical composition, lava is divided into basic, intermediate and acidic. Basic lavas (SiO_2 less than 52 percent by weight) are black, opaque and can contain millimeter-scale crystals of minerals like olivine, pyroxene and plagioclase. Such lava may also contain fragments of rock from the mantle (peridotitic nodules). Intermediate lava (SiO_2 between 52 percent and 62 percent by weight) varies in color from gray to black and almost always contains a large quantity of crystals of plagioclase and lesser quantities of pyroxene, olivine, sanidine and biotite. Acidic lavas (SiO_2 greater than 62 percent by weight) are usually gray and contain a great deal of glass and, in particular conditions, crystals of sanidine and quartz.

On the basis of their crystal content, lavas are classified as either porphyritic or aphanitic: the first contain isolated crystals a few millimeters in size, immersed in a glassy or fine-grained matrix; the second has a uniformly fine-grained matrix composed of tiny glass crystals. Obsidian is a particular variety of aphanitic rhyolitic lava composed of a glassy black or dark mass, very sharp, with characteristics of conchoidal fracturing. Perlite is a kind of very fragile volcanic glass that has the particular characteristic of breaking into small spheroids. Because of the high quantity of water dissolved in glass, perlite, brought to a high temperature, expands to form a sort

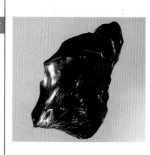

Above: Obsidian is a variety of black or dark volcanic glass, most often composed of rhyolite. It has typical conchoical fractures.

of artificial pumice. It is used industrially as an insulating material for heat and sound.

LAVA FLOWS AND LAVA DOMES

The eruption of basic (basaltic) magma leads to long, thin, fluid lava flows, covering dozens or hundreds of miles. Such basaltic flows are divided into two types that have been given names based on Hawaiian words: those with smooth, ropy surfaces are known as pahoehoe,

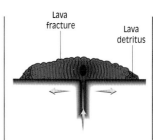

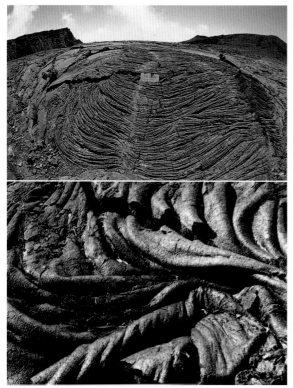

Above: Diagram of the formation of a lava dome caused by the emission of viscous lava that cannot flow away from the eruptive vent. The surface of the lava dome cools when it comes in contact with the atmosphere, fractures, and in some cases produces accumulations of detritus on the slopes of the dome.
Left: Flow of pahoehoe-type lava in Iceland; note the ropy surface.
Below left: Detail of pahoehoe lava in which the rope structure is clearly visible.
Bottom: Flow of a'a lava in Iceland.

those with rough surfaces, jagged or spiny, are known as a'a. When the effusive eruption of such magmas goes on for weeks or months, the flow tends to form a rigid external crust beneath which the flow of molten material continues. When the flow of lava from the vent stops, the molten lava drains away, leaving behind complex underground cavities (lava tubes) that sometimes cover many miles of distance. The flows produced by magma of intermediate viscosity

Right: River of basaltic lava in Hawaii. The central, darker part of the flow is in the initial stages of the formation of a crust.

Below: Hornito composed of carbonatite lava at Ol Doinyo Lengai in Tanzania. This type of structure forms through the accumulation of lava directly over a vent.

Bottom: Opening in the roof of a lava tube revealing the flow of fluid lava inside.

often become covered in angular blocks of lava known as block lava. Such lava often has a thickness of dozens of feet and is extremely common in andesitic and dacitic flows from subduction-zone volcanoes. When such high-viscosity lava pools near the vent it can form domes.

Many domes are steadily fed with abundant new lava, thus tending to "swell" with this progressive addition of lava, in the end forming a single mass of lava (endogenous domes). The outer surface of the dome, exposed to air, cools and hardens, so further expansion from within causes this hardened crust to shatter, spilling detritus down the slopes of the dome to accumulate at its base. When the quantity of lava inside the dome is too great, the outer structure can break, spilling molten lava from the top of the dome, forming a lobe of lava that covers the preexisting mass. Domes of this type are called exogenous. When the lava that emerges from the volcanic conduit is highly viscous, it can behave like a solid mass that does not

flow sideways but rises vertically until it collapses, producing an avalanche of molten rock. The structures that are formed following this type of activity are very steep and can evolve into true rocky spires called spines.

Above left: The spectacular columnar basalt cliffs with waterfall at the Svartifoss in Iceland's Skaftafell national park.
Above: Detail of the basalt cliffs, which formed through the slow cooling of thick lava flows. The flows contracted as they cooled, fracturing at 120-degree angles and forming hexagonal columns.
Left: Aerial view of an andesitic lava dome that formed over the course of the 1980s inside the crater of Mount St. Helens in the Cascades of Washington State. The dome is about 820 feet (250 m) high.

UNDERSEA EFFUSIVE ACTIVITY

Most of the volcanic eruptions that take place underwater are effusive since the pressure of the water prevents or drastically limits the liberation and expansion of gas necessary for explosive activity. The type of lava structure produced in this environment is known as

pillow lava. The "pillows" are roundish masses of lava with a diameter of several inches to several feet, covered by a glassy crust that forms from the sudden cooling of the lava when it comes in contact with the water. When an eruption takes place on the sea bottom, pillow lavas are quickly formed and pile up one atop another to create a mound around the vent. The rapid cooling of the lava caused by contact with the water, combined with the eventual expansion of each pillow, produces the shattering of the glassy crust and the formation of large quantities of glass shards (hyaloclastites), which accumulate in the spaces between the pillows.

Hyaloclastite deposits generated underwater are often transformed into a yellowish material of sandy composition rich in iron oxides called palagonite.

Top: Pillow lavas.
Bottom: Block of palagonite in Iceland.

EXPLOSIVE ACTIVITY

Volcanic magma is made explosive by the violent expansion of gas dissolved in the magmatic liquid; as the magma rises toward the surface the pressure diminishes, releasing the gas. The many types of explosive volcanic activity range from a single discharge of molten fragments several feet into the air to the formation of gigantic eruption columns that reach heights of more than 30 miles (48 km). A particular category of explosive volcanic activity, one that is very spectacular and violent, involves the interaction of magma and water.

FACTORS CONTROLLING EXPLOSIVE ACTIVITY

The principal factors controlling explosive volcanic activity are gas content and viscosity. A third factor of great importance is the quantity of mass, meaning the amount of eruptive material that the volcano discharges over a given period of time. At great depths inside the Earth, volatile substances (volcanic gases) are dissolved in the magma. The rise of the molten mass toward the surface leads to a gradual reduction in pressure, with the consequent liberation of the volatile elements, principally water and carbon dioxide, which form gas bubbles throughout the liquid. As the magma approaches the surface these bubbles occupy a progressively larger volume of the magmatic fluid, and eventually they fuse to form larger bubbles. The process of liberation and expansion of gas is contrasted by the resistance that the magma opposes to the growth of bubbles, with more resistance the more viscous the magma. Chemically evolved magmas, those rich in silica and therefore viscous, effectively oppose the expansion of the gas; as a result, the bubbles preserve a greater pres-

sure inside themselves than that of the surrounding magmatic liquid. Furthermore the process of liberation of the gas can produce an increase in the viscosity of the magma by a factor of hundreds of thousands of times.

When the internal pressure of the gas bubbles becomes too high compared to that of the surrounding liquid magma, or when the viscosity of the magma becomes too high to permit the magma to flow along the conduit, the magma fragments, transforming from a liquid containing gas bubbles to a gas containing fragments of bubbly liquid (pyroclasts).

The depth at which the initial liberation of gas occurs is known as the exsolution level. The exsolution level of the gas and the fragmentation level of the magma divide the magma-chamber-conduit system into three parts: a lower part, formed by magma

Above: Eruptive column formed during the May 1980 eruption of Mount St. Helens in Washington State. Part of the cloud of gas and volcanic ash was generated by the top of a pyroclastic flow, visible at the base of the column itself.

without gas bubbles; an intermediate part, formed by magma containing gas bubbles that occupy a fraction of the upward-moving volume; and finally a portion located above the fragmentation level, where fragments of magma are transported by the rush of the gas. The expansion of the volatile fraction in the conduit is accompanied by a progressive acceleration that eventually reaches the speed of sound. When the mixture of gas and fragments is discharged from the conduit, it can undergo a further acceleration to supersonic speed in the crater.

Inside the conduit, the column of magma assumes different shapes and explosive behaviors according to its viscosity and the quantity of available gas. Magma of little viscosity that is low in volatile substances gives off intermittent explosions of large gas bubbles, each individual explosion leading to the expulsion of a modest amount of material. Viscous magma rich in gas will cause high-velocity discharges of materials from the vent with an explosive regime involving the continuous expulsion of gas and fragments.

Right: The decrease in pressure that accompanies the rise of magma causes the release of some of the volatile elements (magmatic gases) dissolved in the magma, leading to the formation of gas bubbles. The proportion of gas bubbles increases along the eruptive conduit until the level at which the magma fragments. Above the fragmentation level the magma is transformed into a mixture of gas and pyroclasts. The magma accelerates along the eruptive conduit until reaching, in many cases, supersonic velocity as it leaves the crater or vent.

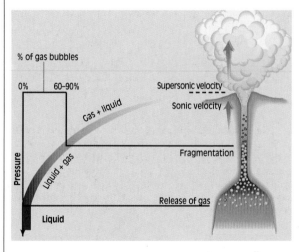

CLASSIFICATION OF EXPLOSIVE ERUPTIONS

Volcanic eruptions are classified on the basis of the level of fragmentation of the magma and the height of the eruptive cloud. Although the spectrum of types has been divided into a practically endless variety, explosive eruptions can be divided into seven basic types: Hawaiian, Strombolian, Subplinian, Plinian, Vulcanian, Surtseyan and Phreatoplinian.

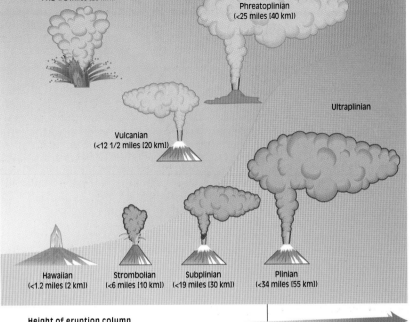

Fragmentation of magma

Surtseyan
(<12 1/2 miles [20 km])

Phreatoplinian
(<25 miles [40 km])

Ultraplinian

Vulcanian
(<12 1/2 miles [20 km])

Hawaiian
(<1.2 miles [2 km])

Strombolian
(<6 miles [10 km])

Subplinian
(<19 miles [30 km])

Plinian
(<34 miles [55 km])

Height of eruption column

Above: Explosive volcanic eruptions divided into those that involve the explosive interaction of magma and water (blue area) and those that do not (orange). On land, the level of magma fragmentation is proportional to the energy released during an eruption. The higher the level of fragmentation the smaller the average size of the pyroclastic products produced. In explosive activity involving the interaction of magma and water, the level of magma fragmentation is very high because of the additional energy supplied by the hydromagmatic explosions. The height of the eruption column depends on the mass of material emitted during a given unit of time.

The first five types are directly controlled by the liberation of gas initially dissolved in the magma; for this reason they are called, in the narrowest sense, magmatic eruptions. Surtseyan and Phreatoplinian eruptions, in which the explosive characteristic is dominated by the vaporization of external water, whether of subterranean origin (aquifer) or on the surface, merit a separate discussion and will be described further on.

HAWAIIAN ERUPTIONS

Hawaiian eruptive activity, so named because it is characteristic of the Hawaiian Islands, involves the emission of low-viscosity basaltic lava with a low level of gas. Hawaiian eruptions often involve lava lakes and lava fountains. Lava lakes are formed when lava collects in craters; the surface of the lava in the crater will be disturbed by the periodic expulsion of large gas bubbles. The activity of a lava fountain consists in the more or less continuous expulsion of a stream of incandescent lava to heights varying from tens to hundreds of feet. Hawaiian-type lava lakes

Above: Lava fountain.
Above right: Birth of the island of Surtsey off the southern coast of Iceland in November 1963. The particular type of volcanic activity involving the explosive interaction of basaltic magma and water is now known as Surtseyan, named after this event.
Right: Gas being released from the surface of a lava lake at Nyiragongo, Congo.

are very rare; active examples include those at Erta Ale in Ethiopia, Nyiragongo in Congo and Kilauea on the island of Hawaii. The material emitted by Hawaiian lava fountains falls to the ground while still in the incandescent state and thus can join to create lava flows. In Hawaii there are also fissure-type eruptions in which magma rises along a fissure and erupts out of several vents at the same time. The most spectacular

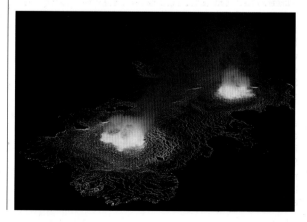

example of a fissural eruption was that of the Laki volcano in Iceland in 1783 when a series of fissures more than 17½ miles (27 km) long opened, from which more than 3⅓ cubic miles (14 km³) of lava flowed, eventually covering an area of about 195 square miles (500 km²).

An example of a fissure-type eruption appears in the photograph on page 13.

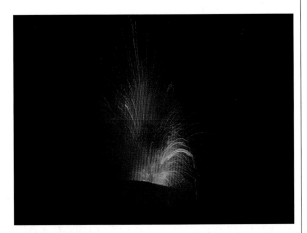

Left: Nocturnal Strombolian activity at Mount Etna on Sicily.
Below: Arenal in Costa Rica is one of the few volcanoes in the world in a perennial state of Strombolian activity. Others include the Italian volcano Stromboli itself, which gives its name to the type of volcanic activity, Pacaya in Guatemala and Sangay in Ecuador.

STROMBOLIAN ERUPTIONS

Strombolian-type eruptions are named for the Stromboli volcano of the Aeolian Islands. These eruptions are produced by medium-low viscosity magma with a gas content generally greater than that of Hawaiian eruptions. The explosions, which follow one another at regular intervals, take place following the explosion of a large gas bubble and lead to the discharge of streams of incandescent lava into the atmosphere, up to hundreds of feet high, and at initial speeds up to 655 feet (200 m) a second. The large bubbles are thought to be produced by the progressive aggregation of small gas bubbles inside the column of magma. Once a large enough bubble has been formed, it rises to the bottom of the crater where it comes in contact with the mass of cooler, more viscous lava. The cooler lava opposes the further rise of the bubble, making it explode. Since a certain amount of time is necessary for a large bubble to take shape, Strombolian eruptions usually have an intermittent character.

The lava that flies into the air in long, arcing jets cools in flight and falls to the ground as scoria, a dark volcanic rock with vesicular cavities caused by bubbles of trapped gas. Notable examples of Strombolian eruptions, aside from those that have been taking place at Stromboli itself since earliest times, have taken place at Mount Etna, most of all during the initial phases of eruptions. Numerous Strombolian eruptions took place at Vesuvius in the period following the major eruption of 1631. When this type of eruption is accompanied by a steady rise of abundant magma, the explosions follow one another more closely and the accumulation of scoria forms a cone; at the same time the degassed magma overflows, forming a lava flow. One of the most famous examples of a Strombolian eruption was the eccentric one of Mount Etna in 1669 that led to the creation of the Monti Rossi cinder cones and the lava flow that devastated the city of Catania.

VULCANIAN ERUPTIONS

Vulcanian eruptions, less frequent than the other types, are usually produced by an explosion of magmatic gas, although in some cases there is the possibility of the limited involvement of external water. Vulcanian eruptions are named after the eruption of the Fossa crater on the island of Vulcano in 1888-90. They are characterized by repeated explosions of medium-high energy with a discharge of large bombs that fall with ballistic trajectory up to several miles from the vent, and by the production of clouds of gas and cinders that rise through convection to a height of around 12 miles (20 km). The magma produced by this type of activity is very viscous and contains a modest quantity of gas.

Below: Seven historical eruptions with data on their important characteristics.

Important historical eruptions	Height of eruption column (km)	Duration in hours	Volume of magma (km³)	Amount of emission (kg/s)
Santorini 1500 B.C.	36	?	?	2.5×10^8
Vesuvius A.D. 79	32	9.5	2.1	1.5×10^8
Vesuvius 1631	19	11	0.2	$1.5 \times 10^7 - 3 \times 10^7$
Tambora 1815	44	2.2	1.2	3.8×10^8
Santa Maria 1902	34	35	8.6	1.7×10^8
Quizapu 1932	30	18	4.3	1.5×10^8
Pinatubo 1991	40	6-9	5	4.2×10^8

PLINIAN AND SUBPLINIAN ERUPTIONS

Plinian eruptions, far and away the most violent and voluminous among magmatic eruptions, are named for the Roman naturalist Pliny the Elder, who died in the famous eruption of Vesuvius in A.D. 79, which destroyed the cities of Pompeii, Herculaneum and Stabia. This type of eruption is produced by viscous magma rich in gas and is characterized by a continuous process of fragmentation of the magma and discharge into the atmosphere of gas and pyroclasts (fragments of magma that when cool form volcanic pumice and ash). The process of magma fragmentation during Plinian and Subplinian eruptions

Left: Explosive eruption at Ruapehu, New Zealand. The interaction between the jet of gas and pyroclasts and the atmosphere causes the formation of vortices in the eruption column that draw in air. As more and more air is drawn in and warmed, the average density of the eruption column diminishes, and if it drops below that of the surrounding atmosphere it will rise, sometimes reaching dozens of miles into the air.

(the only difference between the two is the amount of energy involved in the eruption) produces a large quantity of fine material. During a Plinian eruption large quantities of ash and cinders are discharged from a central vent; the magma flow during these events is estimated to vary between 22 million and 22 billion pounds (10 million and 10 billion kg) per second, with the overall volume of magma discharged often reaching a few cubic miles. The events of A.D. 79, considered the paradigm of Plinian eruptions, threw into the atmosphere a volume of magma equal to half a cubic mile (2.1 km^3) in a period of less than 10 hours with an average flow of around 10^8 kilograms a second. The table on the opposite page gives the dates of some of the most important Plinian eruptions in recorded history.

THE PLINIAN ERUPTION COLUMN: CONVECTIVE CLOUDS AND COLLAPSING CLOUDS

Plinian eruptions form towering eruption columns that rise high into the atmosphere, assuming a shape that somewhat resembles that of a pine tree with cauliflower-shaped clouds. A Plinian column lifts enormous quantities of fragments (up to a few inches in size) to heights of dozens of miles. From the point of view of internal dynamics, Plinian columns are divided into three regions: the gas-thrust region, the convective-thrust region and the umbrella region. In the region of the gas thrust, located immediately above the crater or vent, the mixture of gas and magma fragments accelerated from the

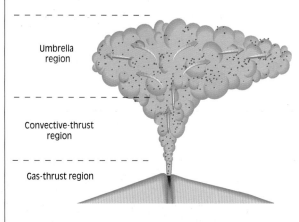

Right: The main regions of a Plinian eruption column. Near the crater or vent the mixture of gas and pyroclasts is shot upward (gas-thrust region); as the initial thrust loses its power the vertical speed of the mixture diminishes; it also loses much of its heavier pyroclastic components. If the cloud of hot gas and particles becomes less dense than the surrounding atmosphere, it will rise through buoyancy (convective-thrust region). The mixture will rise until it reaches an altitude at which it is of the same density as the surrounding atmosphere; at that point (umbrella region) it begins to spread out laterally while losing part of its contents of volcanic ash.

Below: If after the initial upward thrust wears off the mixture is denser than the surrounding atmosphere, it will fall back on its own weight, perhaps to generate a pyroclastic surge that will spread outward along the slopes of the volcano.

Umbrella region

Convective-thrust region

Gas-thrust region

expansion of the gas in the conduit, is shot into the atmosphere at high speed. The height of the area of thrust varies from a few hundred yards to several miles.

In the area of convective thrust, the volcano cloud rises into the atmosphere because it is less dense than the surrounding air. The low density of the cloud results from its high temperature and its continuous appropriation of air, which is warmed by the heat given off by the fragments of magma. The difference in density provides the force to move the cloud through the troposphere (which extends outward from the Earth's surface an average of 7 miles [12 km]) and in many

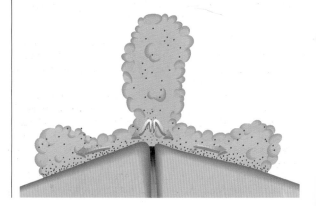

cases to reach the stratosphere, rising to maximum heights of 31 to 34 miles (50 to 55 km). The process of rising by convection ends when the cloud, by then chilled and weighted by the solid materials inside it, reaches the same density as the surrounding air and begins to spread laterally. This is the region of the umbrella, which can extend over hundreds of miles. A thick rain of pumice, rock fragments and sandy material falls from the sides of the column and the umbrella.

A Plinian eruption column does not necessarily involve a convective cloud; the cloud can instead collapse, the mixture of gas and fragments reaching an apex only to fall back to the ground. This happens when the forces lifting the mass are exhausted, the mixture is not sufficiently diluted or it has an average density greater than that

Below: Cloud of volcanic material at Santiaguito, Guatemala. In this case the column of gas and volcanic ash does not form over a vent but lifts off the top of a moving pyroclastic flow that is giving off warm gas lighter than the surrounding atmosphere. An eruptive cloud of this kind is called a co-ignimbrite (or Phoenix) column. A co-ignimbrite column behaves much like a Plinian column except of course that it does not involve a region of gas thrust. In the

of the surrounding air. The gravitational collapse of the cloud can lead to the formation of a high-speed lateral avalanche of enormous quantities of gas, dust and hot fragments of magma. Such a turbulent rush is called a pyroclastic flow.

Pyroclastic flows produced by Plinian eruptions can expand radially around the volcano, in exceptional cases covering distances of over a hundred miles. During the flow, solid or liquid materials suspended in the gas tend to concentrate toward the bottom, forming a stream of dense particles that concentrate on lower areas. At the same time, the hot gas, lighter than the air, separates out, moving upward and rising off the flow to create secondary eruption clouds (co-ignimbrite clouds) that can transport enormous quantities of ash.

region of the umbrella, which develops at heights that vary from a couple to a few dozen miles, the expansion of the cloud and the dispersion of the volcanic ash are affected by atmospheric winds.

Above: Maurice Krafft, famous photographer and maker of documentaries on volcanoes and volcanic eruptions, while shooting a pyroclastic flow generated by Pelean activity on Augustine in Alaska. Krafft died on June 3, 1991, together with his wife, Katia, and another 30 people while filming similar pyroclastic flows at Unzen in Japan.

Right: The three types of phenomena that generate the flows of gas and fragmentary material known as nuées ardentes.

PELEAN ERUPTIONS
The explosive activity of domes

Eruptions that cause the formation of avalanches of incandescent material associated with the formation of domes are collectively called Pelean eruptions—named after Mount Pelée, where this phenomenon was first described in 1902. Lava domes are usually produced by the extrusion of intermediate and acidic magma (andesitic, dacitic, rhyolitic) low in gas. Because of their tendency to form very steep sides, such domes are subject to collapsing in avalanches, an event often associated with explosions. The explosion of a lava dome differs from the explosion of magma from a volcanic vent since the lava is not directed vertically and involves the sudden discharge of large quantities of material. For the sake of simplicity these events are divided into three types, all three of which result in the formation of flows of incandescent material and gas (known as nuées ardentes, "glowing clouds") that behave much like the pyroclastic flows produced by Plinian columns.

A. *An avalanche of small portions of the body of the dome*
B. *An avalanche of portions of the dome accompanied by a small explosive event*
C. *The explosion of the entire mass of the dome, usually following a large avalanche*

The first type (A) involves the formation of an avalanche of incandescent blocks and cinders that moves at relatively modest speeds down depressions cut in the volcanic cone. Merapi in Indonesia (see Number 34, pages 178–79) is among the best known for this activity.

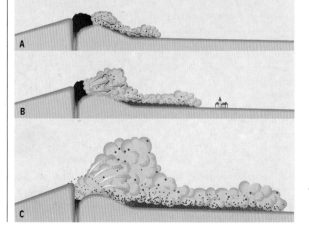

In the last type (C), the sudden explosion of the entire mass of the dome produces an eruptive cloud denser than the air that moves at high speed, transporting a turbulent mass of material. In some situations, the eruptive cloud moves radially around the volcano (for example, in the eruption of Lamington in Papua New Guinea; see Number 31, pages 172–73), while in others it is projected laterally, thus striking a single area (as in the lateral explosion of Mount Pelée on Martinique in 1902, Bezymianny in Kamchatka in 1956, and St. Helens in Washington in 1980; see numbers 99, 48 and 56). In the last two events, the mass of lava was enclosed in the volcanic cone (cryptodome formation), and the explosion involved the collapse of part of the volcanic cone itself. Such lateral blasts are known for their violence and destructive power, often traveling distances of 9 to 12 miles (15 to 20 km) from the volcano, destroying any and all structures that happen to be in their path. The second type (B) is intermediate since the event of the avalanche is joined by a mass of additional material, produced by an explosion of the internal, hotter portion of the dome exposed by the avalanche. During its movement, the rough material concentrates rapidly toward the bottom, forming an avalanche similar to those produced in type A; the part of the cloud containing fine material in suspension strikes the nearby

areas of the bottom of the valley, thus increasing the area devastated. A notable example of this type of activity was the explosion of Mount Unzen (see Number 43, pages 198–99) in Japan on June 3, 1991, during which the French volcanologists Maurice and Katia Krafft and the American Harry Glicken lost their lives. The explosion of domes can set off a regressive fragmentation of the magmatic mass inside the conduit with the eventual development of sustained explosive activity similar to that of Plinian eruptions. Normally such explosive phases are brief and are followed by the extrusion of a new dome.

Below: Pyroclastic flow generated by Pelean activity on Augustine in Alaska. Photographed by Krafft.

Below right: Large lava bomb on the edge of Fossa crater on Vulcano, one of the Aeolian Islands, Italy. Below, from top to bottom: Pair of fusiform, or spindle, bombs; bread-crust bomb; pumice. The cracks in the breadcrust bomb are produced by the release of gas inside the bomb, which increases the volume of the bomb, causing the cracks.

THE PRODUCTS OF EXPLOSIVE VOLCANIC ACTIVITY

Explosive volcanic activity usually involves the vertical discharge of a jet of gas and incandescent fragments of lava. Such explosions also hurl outward fragments of solid rock torn from the walls of the crater or the conduit (lithic deposits), as well as pieces of crystallized rock from the walls of the magma chamber (lithic magma). Such fragmented material formed by an explosion or ejected from a vent is collectively called pyroclastic material or tephra. Pyroclastic material

Particle size in millimeters	Pyroclastic fragments	
> 256	Coarse	
		BLOCKS AND BOMBS
64–256	Fine	
2–64		LAPILLI
0.063–2	Coarse	
		ASH
0.004–0.063	Fine	
< 0.004		

that hardens to form rocks is called tuff. Some tuff is soft enough to be dug up and has been used since antiquity as a thermal or acoustic insulation. Pyroclastic material can also be divided into groups according to size, type and internal structure.

Left: Stratified deposits of volcanic ash. The stratification is produced by the successive depositing of layers of volcanic material and can be "read" by volcanologists to reconstruct eruptive events.
Below: Block of tuff, a pyroclastic material mixed with abundant quantities of volcanic ash.
Bottom: Several products of effusive basaltic activity. At the upper left is Pele's hair (a kind of spun volcanic glass); to the right are Pele's tears (drops of volcanic glass). Both are named for Pele, the Hawaiian goddess of volcanoes. At the bottom left is reticulite, also called thread-lace scoria (a feathery pumice). All are products related to the release of magmatic gas from extremely fluid magma.

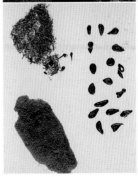

DEPOSITS OF EXPLOSIVE ACTIVITY

The fragmented materials ejected from the vent of a volcano undergo various processes of transport before being deposited. According to the mechanism of transport and dispersion, the pyroclastic deposits are divided into three basic categories: fall (deposits that drop from the sky), pyroclastic flow (those brought by flow) and surge deposits.

FALL DEPOSITS

The fragments ejected into the atmosphere or lifted high into the air by the convective currents in an eruption column eventually fall to the ground due to gravity. The larger fragments fall at high speed in the immediate area of the vent, following parabolic trajectories, while the smaller pieces, those only an inch or two in size, are lifted by gas and carried away from the area of the vent. The type of material shot into the air to then fall back depends on the type of explosive activity. Hawaiian and Strombolian eruptions project material to modest height, producing principally rough deposits of scoria and

lapilli (the plural of *lapillus*, meaning "little stone," these are small stony or glassy fragments of lava). Most of this material falls back around the area of the vent, leading to the construction of cinder cones. In the course of Plinian eruptions finer material (ash) and light, spongy fragments (pumice) can be swept up into the eruption column to reach heights of many miles.

Once the cloud reaches its maximum height, it expands laterally (forming the umbrella) and the fragments that have been swept up into the column begin to fall, creating a thick rain of material. In the course of Plinian eruptions the rain of material can go on for hours and even days, causing an accumulation of layers of lapilli and ash that cover surfaces over hundreds or thousands of square miles like a snow-fall. The major characteristics of the fall of Plinian deposits are their relatively consistent thickness and the fact that they are constituted of fragmented rock material—pumice—that tends to be uniform in size.

Top: Diagram showing the process of ashfall deposits.
Above: Plinian column in 1991 from Pinatubo in the Philippines. The column reached a height of more than 25 miles (40 km), producing an enormous fall of volcanic ash and cinders over an area of more than 74 million acres (300,000 km²).
Above right: Ashfall deposits cover surfaces with layers of the same thickness.

DEPOSITS OF PYROCLASTIC FLOWS

Pyroclastic flows are produced by eruptions, by the collapse of eruption columns or by the explosive activity of lava domes. The flow is a high-density mixture of gas, ash and fragments of incandescent lava (pyroclasts) moving down the slope of a volcano at high speed under the action of gravity. Because of their high density these flows adapt to the topography they move across. As the slope diminishes, the flow moves more slowly, finally coming to a halt, filling depressions and forming deposits. The deposits left by a pyroclastic flow are a chaotic mixture of cinders, ash and volcanic materials, including fragments of various size; quite often such deposits also contain carbonized plant material carried along by the flow. If the temperature of the flow is hot enough when it comes to rest, the pyroclasts may

meld and solidify, forming a mass so compact it will resemble the result of a lava flow.

SURGE DEPOSITS

A pyroclastic surge is less dense than a pyroclastic flow because it contains more gas; a surge is a ring-shaped cloud of gas and solid debris that rushes radially away from a volcano at high speed. Surges begin at the base of an eruption column; the term *surge* is meant to recall the ring that develops at the base of a nuclear cloud. Because they are denser than the surrounding atmosphere, surge clouds move close to the ground, sweeping away most of what they encounter. Because they move at such high speeds, their direction is less affected by topographic factors; while most surge deposits are found at the bottom of valleys, they also show up in thinner layers on elevated spots. During the surge, the solid fragments in the cloud are held in suspension by the turbulence of the mixture, but as the surge

Above left and above: Deposits of material from pyroclastic flows fill depressions, leaving the terrain flat. Below left and below: Surge deposits. Such deposits tend to blanket surfaces, creating important variations in thickness between areas that are low or high.

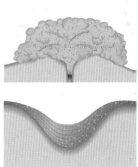

loses speed, the larger, heavier fragments concentrate at the lower levels before finally coming to rest on the ground. An important characteristic of the deposits left by pyroclastic surges is their marked internal stratification and the development of a wavy structure that recalls sand dunes created by the action of wind (see the photo at the bottom of page 61). Compared to the deposits left by flows, those left by a surge have fewer fine particles; the particle size of a surge falls midway between that of volcanic fall and a flow. The clasts in a surge deposit are often partially rounded, a result of the violent abrading encountered during transport. The explosions that produce a surge spread radially around the volcano and often throw blocks of rock that, when falling, leave depressions in the layers of ash and lapilli (impact structures).

IGNIMBRITE
Regional deposits of pyroclastic flows

Ignimbrite (from two words meaning fire and rain) is a rock as compact as lava with a particular structure in which dark, lens-shaped elements (flames) stand out against a lighter background matrix. The flames are the remains of the original pumice, which while still in a plastic state was flattened by the pressure of other material falling onto it. Ignimbrites, also called ash flow tuff or welded tuff, constitute a particular category of pyroclastic-flow deposit. Ignimbrite deposits typically cover very large areas and are produced by eruptions that produce great quantities of magma—tens, hundreds, even thousands of cubic miles—over a relatively short time. Eruptions involving less volume usually produce pyroclastic flows whose movement is more affected by topography, such that the deposits are concentrated in valleys. Larger-scale explosive eruptions generate pyroclastic flows that move at high speed, spread out radially from the vent, and are far less affected by topographical factors, so much so that even at dozens of miles from the volcano they can still move over rises 3,300 feet (1,000 m) high. Thus although

Above: Ignimbrite in the Valley of Ten Thousand Smokes in Alaska, produced by the eruption of Novarupta in 1912. The total volume of ignimbrite is about 6 cubic miles (25 km³), with thicknesses at some points of more than 165 feet (50 m).

most ignimbrite deposits are found inside valleys, they also show up along the slopes of hills. *The typical flames can form only when the volcanic material is still at a high temperature when deposited; thus it often happens that ignimbrite deposits nearest the eruptive vent have the flame structure, while those farther away are composed of normal pumice and ash. The large tuff deposits in the volcanic areas of Latium and Campagna in Italy, around lakes Bolsena and Vico Bracciano, and at Campi Flegrei (see Number 1, pages 104–105) are famous examples of regional deposits of ignimbrite produced by eruptions that took place during the last 500,000 years.*

Left: Fumarolic pipes visible inside a deposit from a pyroclastic flow. Such structures form following the violent release of gas trapped in a deposit. As the gas rises it drags fine material with it, leaving behind a vertical tube as indication of its passage that has a particle-size measurement greater than that of the rest of the deposit. Such pipes vary in length from a few inches to several feet.
Below: Microscopic image of a cross-section of ignimbrite. The darker matrix is composed of welded fine ash. The lighter lenticular shapes that form the flames are remains of pumice that was deformed by the weight of the deposit and by the high heat that prevailed during the entire process of deposition.

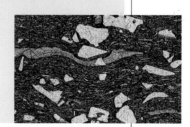

ERUPTIONS WITH THE INTERACTION OF WATER AND THEIR PRODUCTS

During its rise to the surface, magma can come in contact with areas of rock saturated with water or with areas of surface water. In some cases, the contact of high-temperature magma and water causes a large explosion, the result of the process of instant evaporation. The transformation from liquid water to steam causes an enormous increase in atmospheric pressure (an explosive expansion); the energy produced by such explosions causes further fragmentation of the molten magma.

Explosive activity caused by the interaction of magma and underground water is called phreatomagmatic; such activity caused by the interaction of magma and surface water, such as a shallow sea, lake or glacier, is called hydromagmatic.

In contact with water magma loses heat rapidly. The rapid loss of heat causes quick cooling and leads to the formation of fragile volcanic glass crossed by fissures of contraction produced by the thermal

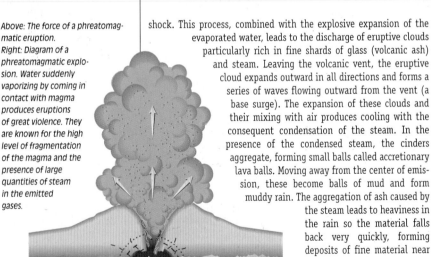

Above: The force of a phreatomagmatic eruption.
Right: Diagram of a phreatomagmatic explosion. Water suddenly vaporizing by coming in contact with magma produces eruptions of great violence. They are known for the high level of fragmentation of the magma and the presence of large quantities of steam in the emitted gases.

Rocks saturated with water

Zone of water/ magma contact

shock. This process, combined with the explosive expansion of the evaporated water, leads to the discharge of eruptive clouds particularly rich in fine shards of glass (volcanic ash) and steam. Leaving the volcanic vent, the eruptive cloud expands outward in all directions and forms a series of waves flowing outward from the vent (a base surge). The expansion of these clouds and their mixing with air produces cooling with the consequent condensation of the steam. In the presence of the condensed steam, the cinders aggregate, forming small balls called accretionary lava balls. Moving away from the center of emission, these become balls of mud and form muddy rain. The aggregation of ash caused by the steam leads to heaviness in the rain so the material falls back very quickly, forming deposits of fine material near the volcano. Laboratory experiments have established that the explosive energy created in the process of interaction between water and magma is greatest when the ratio between the water and magma is 0.1 to 1. When this ratio is less than 0.1 (meaning little available water) the explosive energy abruptly diminishes; when the ratio is greater than 1 the explosive energy gradually diminishes with the increase in the quantity of water.

The table on page 66 gives the different types of activity and structures produced by the different water-to-magma ratios. In order of increasing interaction, the results are cinder cones, tuff rings, tuff cones and deposits of pillow lavas. A cinder cone (a volcanic cone built of fragmented material) involves interaction with a very small amount of external water. Tuff rings are produced by high-energy eruptions and involve the accumulation of debris around a vent located in water, with a water-to-magma ratio of 0.1 to 1.

The energy of the explosions causes the widespread dispersal of volcanic material and the formation of a crater, opened at least partially in rock. A maar is a more or less circular crater occupied by water—a lake, pond or marsh. Tuff cones, usually produced by the interaction of magma and surface water, have smaller craters with steeper sides. When there is a great deal of water (or deep water), volcanic explosions are inhibited and the flow of lava may lead to the formation of pillow lavas. Hydroplinian or phreatoplinian eruptions

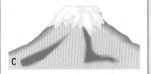

Top: Three different situations that can lead to hydromagmatic eruptions. A) in a crater lake; B) in shallow water; C) with the presence of a summit glacier.
Above: Various types of products from the water/magma interaction. From left to right: Vesicled tuff, accretional yellow lapilli, accretional gray lapilli (pisolites).
Left: Hydromagmatic eruption inside a crater lake at Ruapenu, New Zealand, in 1995.

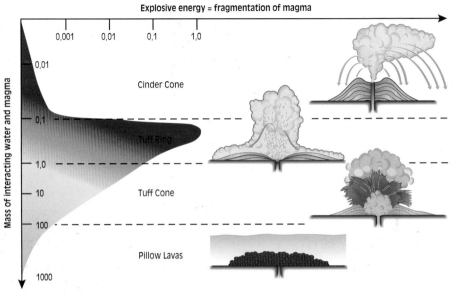

Explosive energy = fragmentation of magma

0,001 0,01 0,1 1,0

Mass of interacting water and magma

0,01

Cinder Cone

0,1

Tuff Ring

1,0

Tuff Cone

10

100

Pillow Lavas

1000

Above: Table indicating the explosive energy of water/magma eruptions from various types of volcanic structures. The maximum explosivity is reached when the relationship between the masses of interacting water and magma is between 0.1 and 1.

Right: The maar of Ukinrek, Alaska, which formed in 1977, has offered scientists their first opportunity to observe the birth of such structures.

occur when the volcanic vent opens in a lake or shallow sea. Once again, the principal effect is the evaporation of large amounts of water followed by condensation and the formation of muddy rain on a large scale.

HYDROTHERMAL ACTIVITY

During the phases of quiescence between eruptions and during eruptions themselves, volcanoes give off large quantities of gas. Almost all volcanic craters are occupied by fumaroles, vents from which vapors and gases are released into the atmosphere (see the figures and tables on pages 30 and 31). The temperature of fumaroles varies from less than 212°F (100°C) to temperatures over 1292°F–1472°F (700°C–800°C), near the temperatures typical of magma. The peripheral areas of a volcano may be the sites of further fumaroles and hot springs of various temperatures containing a variety of minerals. The fluids brought to the surface by fumaroles and hot springs can result from the degassing of magma deep in the Earth or from the warming

Hydrothermal manifestations in volcanic areas.
Above left: The mud on the island of Vulcano is kept hot and fluid by emanations of subterranean gas.
Above: Surface of a dried mud pot crossed by numerous cracks from drying.
Below left: Mud pot, a well of boiling mud in the volcanic area of Lake Myvatn in Iceland.

Above: Hydrothermal spring at Geysir (Iceland). The central hole is the conduit of an extinct geyser. Right: Diagram of the hydrothermal circulation in a volcanic area. The principal component of volcanic gas is water. Some of it comes from degassed magma, but a great deal of it is underground water, the abundance of which is related to precipitation. In volcanic areas the hot subterranean water feeds such surface manifestations as fumaroles, gas emanations and hot springs.

of subterranean water by the magma's heat. In most cases, the fluid has a hybrid origin, being the result of a mixture of the chemical interaction between magmatic gas and underground water.

The temperature increase of the liquid in a subterranean aquifer (caused by the heat given off by magma) can lead to sudden explosions known as phreatic eruptions. The explosive conditions are reached when the liquid in the aquifer, sealed by a rocky cover, reaches a temperature that exceeds the boiling point, generating enough pressure to break the impermeable cover.

Phreatic eruptions produce a jet of mud and pulverized rocks, that have been transformed by the process of hydrothermal alteration, together with quantities of steam. Although they involve small areas of only a few hundred yards, phreatic eruptions constitute a serious danger because they frequently take place without warning. In 1989 an explosion of this type killed nine tourists who were on the rim of Etna's central crater. In this instance, the explosion occurred as a result of small avalanches down the inner slopes of the crater. These avalanches had closed off the vents on the bottom of the crater from which gas had been discharged, leading to a buildup in pressure and finally the explosion. Phreatic eruptions sometimes occur briefly before the reawakening of a dormant volcano. Such was the case with the eruptions of Mount St. Helens in the United States in 1980, of Nevado del Ruiz in Colombia in 1985 and of Pinatubo in the Philippines in 1991.

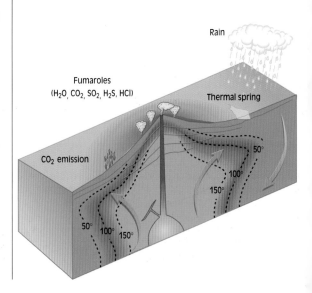

In the case of aquifers heated by magma that are not sealed off by impermeable covers, the heated water will rise to the surface and create various kinds of thermal events. The most spectacular of these are geysers: jets of boiling water, steam and gas that periodically burst from the depths of the Earth. Geysers usually form in the area of a near-extinct volcano where there are underground thermal anomalies and large reserves of groundwater. Famous examples of geysers include Old Faithful in Yellowstone National Park (U.S.A.) and the currently dormant Big Geysir in Iceland, both with jets that reach heights between 100 and 200 feet (30 and 60 m).

Areas of volcanic activity are also characterized by the widespread release of gas (principally carbon dioxide) from the ground, often in quantities greater than are released by fumaroles. The release of carbon dioxide from the bottom of lakes in volcanic craters can produce dangerous concentrations of gas in the water. Under certain conditions this can be suddenly released with the emission of a gas cloud that, being heavier than air, moves along at ground level, suffocating both humans and animals. An event of this type occurred at Lake Nyos (Cameroon) in 1986, killing 1,700 people.

Above: Blue Lagoon at the Svartsengi geothermal power plant near Keflavik, Iceland. The station produces electricity for the capital, Reykjavik; the artificial lagoon is a result in the outflow of water after the heat has been extracted from it. The water of the lagoon is 158°F (70°C); the steam that feeds the station comes out in the ground at 464°F (240°C).

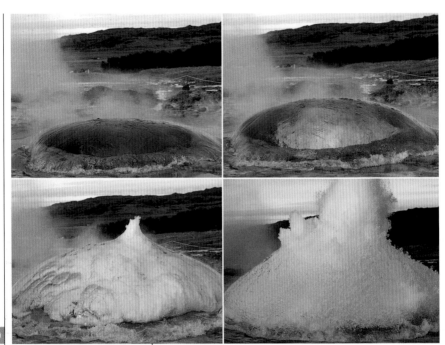

Above: Sequence of photographs showing phases of activity at the Strokkur geyser in Iceland. The surface of the water is agitated; it swells; it explodes; the jet of hot water gushes into the air. The word geyser comes from an Icelandic word meaning "gusher." The activity shown in the four photos takes about one second of time.

LAHARS
Mudflows produced by volcanic activity

Areas of volcanic activity are often the site of floods of volcanic debris and mud. Known by the Indonesian term lahar, *such events are capable of causing more damage and death than volcanic eruptions. Explosive eruptions that cause the accumulation of large masses of ash and pumice on the steep slopes of a volcano are often followed by gigantic debris flows or mudflows capable of destroying areas up to hundreds of miles away from the volcano. The eruption of Nevado del Ruiz in Colombia on November 13, 1985, caused part of the volcano's summit glacier to rapidly melt, producing mudflows that traveled down river valleys over a distance of more than 43 miles (70 km), razing the town of Armero to the ground and killing more than 23,000 people. A similar event occurred in Iceland in October 1996 when the Grimsvotn volcano, located beneath the Vatnajökull glacier, erupted, melting ice from the glacier and leading to the gradual accumulation of an enormous quantity of water; this water created a large underground river equal in capacity to the Rhine,*

70

running 31 miles (50 km) under the glacier to exit to the south toward the ocean, sweeping away roads, bridges and electric lines along the flood plain. The most startling example from this century, however, followed the eruption of Mount Pinatubo on June 15, 1991, in the Philippines. The volcano erupted more than 2½ cubic miles (10 km³) of ash and debris, most of which formed pyroclastic flows down the slopes of the volcano. Torrential rains led to enormous mudflows that swept down the valleys, killing 400 people and leaving nearly half a million homeless. In 1919 the eruption of Kelut, on the island of Java, involved the discharge of about 1.4 billion cubic feet (40 million m³) of water from the lake inside the crater; the mass of water mixed with volcanic ash and swept 19 miles (30 km) downhill, destroying hundreds of villages and killing 5,000 people. Efforts were then made to drain the crater late, and after several attempts the level of the lake was substantially lowered, reducing the risk of a lahar. Lahars produced as a result of explosions inside crater lakes occurred in New Zealand's Ruapehu volcano. Large-scale floods can be caused by the deposit of ash from phreatomagmatic eruptions. Such ash is impermeable and keeps water from filtering into the ground, thus leading to flooding.

Below: The lahar that formed following the eruption of the Grimsvotn volcano in Iceland in October 1996. The heat produced by the eruption caused the partial melting of the glacier, creating a river that flowed at a rate of hundreds of thousands of cubic feet per second and covered thirty-odd miles below the glacier before flowing onto the surface. In Icelandic, this phenomenon is called a jökulhlaup.

Below: From left to right: Small mud volcano, a few inches high, at Nirano (Modena, Italy); small mudflow at Nirano; the Strokkur geyser in Iceland at the height of its activity. Its jets reach about 100 feet (30 m) in height.

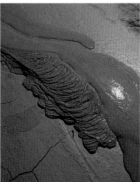

VOLCANIC STRUCTURES

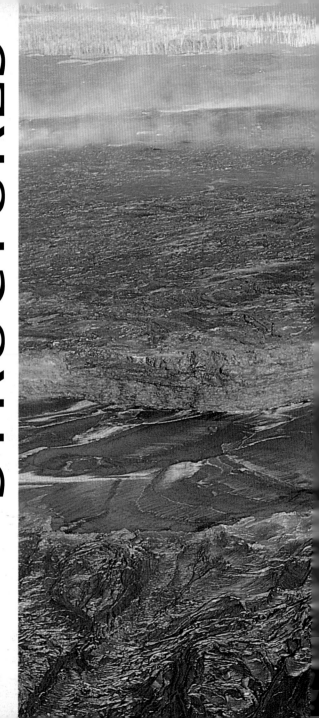

The continuous accumulation of material erupted near a vent causes the formation and progressive growth of a volcanic structure. Since they result from the interaction of various processes, volcanic structures take very variable forms. Some phenomena, such as the extrusion of lava domes, Strombolian eruptions and a series of lava flows discharged from the same vent, result in the ongoing, systematic construction of regular volcanic cones. Discharge of material from another vent, as often happens with volcanoes, will lead to the aggregation of several structures or to radical changes in the normal, regular growth. Other processes, such as large eruptions, the collapse of calderas, avalanches or erosion, tend to destroy volcanic structures or to profoundly modify their morphology.

VOLCANIC CONES

There are two basic kinds of cones: those produced by a single eruption, known as monogenetic, and those formed by the progressive accumulation of eruptive material emitted from the same vent, which are called polygenetic. Monogenetic cones produced by eruptions that do not involve external water are further divided into cinder cones and spatter cones. The first (also known as scoria cones) are created by the fall of partially cooled loose fragments that form around a vent. The latter are formed through the accumulation of strips of lava that melt together as they are deposited.

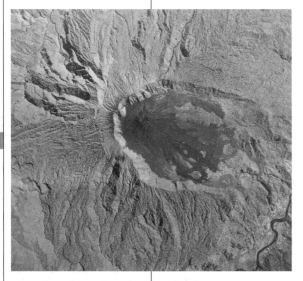

Above: Aerial photograph of the horseshoe-shaped depression atop the Reventador volcano in Ecuador. The structure formed following an enormous avalanche that collapsed the summit of the volcano. Visible inside the depression is the current cone of the volcano. The structure measures roughly 2½ miles (4 km) from north to south.

74

Polygenetic central volcanoes in which the activity always takes place from the same vent are divided into shield volcanoes and stratovolcanoes. Shield volcanoes are structures with very gentle slopes, almost entirely formed by flows of fluid lava, very extensive but of limited thickness. The most typical examples of shield volcanoes are those of the Hawaiian Islands. The volcano of Mauna Loa, for example, rises from a depth of about 14,750 feet (4,500 m) below sea level to a height of 13,681 feet (4,170 m) above sea level, making it the highest mountain on our planet, with a base of nearly 31 by 62 miles (50 by 100 km) and an average incline of less than 10 degrees. The size of this volcano is such that the weight of it produces a sort of sag in the lithosphere. Geophysical studies have shown that a large conduit occupied by molten material is constantly present inside Mauna Loa; research into the composition of the magma erupted suggests that the process of crystallization goes on inside the conduit itself. This process leads to the accumulation of enormous quantities of crystals, primarily olivine, which being denser than the magma liquid are concentrated in the lowest part of the conduit.

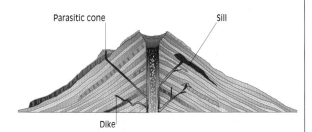

Parasitic cone · Sill · Dike

Stratovolcanoes are conical structures that rise majestically off plateaus along the active continental margin or as island arcs. Viewed in profile, stratovolcanoes can be seen to be conical, sloping toward the top. The lower flanks are only slightly inclined, while the upper areas commonly reach angles of 30–35 degrees. As indicated by its name, a stratovolcano is typically formed by alternating strata of lava flows and pyroclastic material, erupted from a central crater. The summit is formed of fragments of coarse material, both of explosive origin and produced by lava flows that breached the central crater. The lower slopes are composed of lava flows alternating with pyroclastic material, such as pyroclastic flows and surges, deposits of ashfall, along with layers left by mudflows. The slopes of many stratovolcanoes are sometimes dotted with parasitic cones, the results of the lateral discharge of magma.

Because of the steep angle of their sloping flanks, together with other factors related to their structure, stratovolcanoes are often the site of catastrophic avalanches. In many instances these cause the removal of the topmost part of the cone, leading to the formation of giant breaches, or openings, left by the material breaking away, often

Above left: Cross-section diagram of a stratovolcano. Such edifices are created by the accumulation of layers of magmatic rocks of varying nature, both intrusive (dikes and sills) and effusive (lava flows and pyroclastic deposits of various nature). The terms dike and sill refer to bodies of intrusive magma. A dike has a sheetlike form and cuts across the layering of the rock in which it intrudes; a sill is also tabular but runs parallel to the layers of the rocks it intrudes. When a dike (or more rarely a sill) intersects the surface of the volcano, it produces a lateral eruption with the formation of parasitic cones.
Below: Lateral avalanche of a volcanic structure with the formation of hummocks at the foot of the volcano.

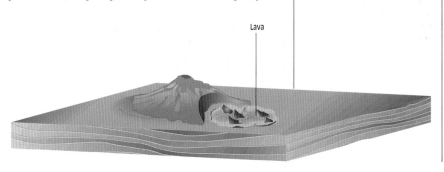

Lava

forming a characteristic horseshoe-shaped summit crater. The material in the avalanche pours down the side of the volcano to spread out at its foot, forming extensive detritus fields marked by many little hills called hummocks. These are exclusive to volcanic landscapes and represent areas of harder rock (generally lava) that rise above the mass of detritus composed of friable material. After the avalanche, activity returns to normal inside the horseshoe-shaped crater, often leading to the construction of a new volcanic cone. There are many examples of volcanoes that have been subject to large-scale avalanches. Among the most famous are Mount St. Helens (U.S.A.), Bezymianny (Kamchatka), Galeras (Colombia) and Reventador (Ecuador). Notable examples in Italy are Sciara del Fuoco in Stromboli and the Valle del Bove on Etna.

Volcanic structures are created by the accumulation of erupted material, but in a certain sense they are also composed of the solid bodies that have formed following the cooling and solidification of intrusive magma inside the volcano. Magma of low viscosity tends to wedge itself into fissures in the rock, resulting in long, narrow bodies that cut across the body of the cone. Structures of this type are called dikes and have different names according to their particular geometry or the processes that led to their formation (see the diagram below). The most commonly encountered category is the radial dike, which results from the fracturing of the cone caused by the pressure of magma in a conduit. In some cases, dikes are arranged equally in different directions; in others, the volcanic structure tends to break along preferential lines and one therefore finds the formation of strips of dikes following the

Above: A dike running perpendicular to rocky layers.
Below right: Dikes are divided into types according to their location within a volcanic edifice.

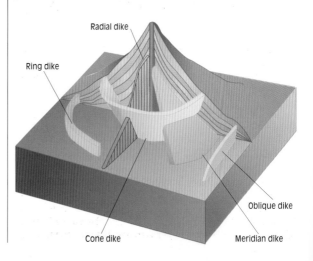

Radial dike

Ring dike

Oblique dike

Cone dike

Meridian dike

structure. Some dikes are the remnants of conduits through which magma was discharged out of the volcano and onto its slopes in lateral eruptions.

CRATERS AND CALDERAS

A volcano is often the site of a circular depression, sometimes occupied by a lake. Such depressions are commonly called volcanic craters. An important distinction must be made between craters and calderas, for they differ in size and in the mechanisms behind their creation. Craters are usually smaller than calderas, most less than 3,280 feet

(1 km) in diameter, and they are the direct result of explosive activity at a volcanic vent along with the collapse or erosion of the walls. Calderas (from the Spanish word for cauldron) are large circular depressions produced by phenomena of subsidence connected to the collapse of the roof of a magma chamber. The collapse of a magma chamber usually results from the loss of internal pressure following a large eruption.

Many calderas have lakes and are sites of extraordinary beauty and fascination. Notable examples of calderas produced in historical or prehistoric time are those of Crater Lake in Oregon, Santorini in the Aegean and Campi Flegrei near Naples. One special example is the caldera of Katmai in Alaska, produced by the eruption of the Novarupta volcano from which ignimbrites were deposited in the Valley of Ten Thousand Smokes. In this case the magma from the magma chamber of Mount Katmai was erupted by a vent located 5 miles

Above: Summit caldera of Pinatubo in the Philippines following the 1991 eruption. Layer upon layer of volcanic ash creates an excellent sealant, thus allowing rainwater to accumulate in hollows and creating caldera lakes.

(8 km) away. The partial emptying of the chamber caused the collapse of the upper part of Katmai, forming a circular depression 3,300 feet (1,000 m) deep with a diameter of 3 miles (4.8 km). The majority of calderas are produced by explosive eruptions that involve a large amount of material.

The sequence of the phenomena is recreated in three basic stages: formation of a Plinian eruption column from a single vent; the beginning of the collapse of the roof of the magma chamber, with the opening of more vents; the collapse and formation of the caldera. It is during the second stage that magma is ejected at great speed from the chamber, producing pyroclastic flows of large extent (ignimbrites).

Above: The church of Saint-Michel d'Aiguille at Le Puy in Auvergne (Massif Central, France), built atop a volcanic neck dating back two million years.

Right: Diagram of the formation of a caldera. During the course of a volcanic eruption the pressure in the magma chamber is diminished, destabilizing the entire volcanic edifice; a series of concentric fractures are formed through which magma is discharged; the structure collapses, leaving a caldera depression.

The collapse of a caldera can happen in two ways, pistonlike and piecemeal, or chaotic. The pistonlike collapse takes place following the lowering of a single cylindrical block of rock; it seems probable that in such cases the lowering of the block is gradual and

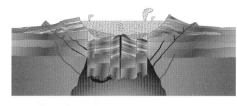

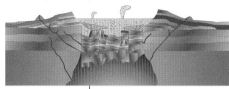

accompanies, at least for a certain time, eruptive phenomena. Chaotic collapses take place following the explosive ejection of blocks of rock that form the roof of the magma chamber, with the disorderly collapse of the remaining blocks.

After their formation, caldera depressions can be the sites of further volcanic activity. The accumulation of eruptive materials from activity following the formation of the caldera can lead to the partial or even total refilling of the caldera, such that in some cases the depression is no longer visible. An example of a caldera almost completely refilled by new volcanic products is that of Campi Flegrei, formed followed the eruptions of the Campanian ignimbrite (36,000 years ago) and the Neapolitan Yellow Tuff (14,000 years ago).

Some calderas are subject to a particular phenomena called resurgence that involves the raising of the caldera floor, driven upward by the flow of new magma from below as it refills the magma chamber. One of the best examples of a resurgent caldera is Monte Epomeo on the island of Ischia, composed of a mass of tuff deposited inside a caldera formed about 55,000 years ago and later raised more than 2,625 feet (800 m).

Above: Two types of caldera collapse, piston-like (top) and chaotic (bottom). Below: Strombolicchio off the island of Stromboli in the Aeolians, Italy, is an example of a volcanic neck.

VOLCANIC NECKS

The removal of a volcanic structure through erosion may lead to the formation of a special kind of structure. In general, erosion causes the carrying away of loose material; since the lava that forms dikes is more resistant over time, as the other material is eroded away the inner volcanic rock remains, forming a relief with very steep slopes.

The term *neck* is used for these towers, which are composed of the lava once inside volcanoes that has been uncovered through erosion. Notable examples are those of Le Puy in Auvergne (France), formed two million years ago, and the Devil's Tower in Wyoming. The reef of Strombolicchio, located to the northeast of the island of Stromboli, is another fine example of a volcanic neck.

VOLCANO MONITORING

*Preceding pages: Sakura-Jima and
the city of Kagoshima on the island
of Kyushu, Japan.*

A volcanic eruption is certainly one of the most awe-inspiring natural
phenomena one can witness. At the same time, of course, such eruptions are among the most catastrophic and destructive events to
occur on our planet. The 1900s alone saw many highly destructive
eruptions in which tens of thousands of people lost their lives and
vast areas, both wild and settled, were devastated: Mount Pelée (Martinique) in 1902, with the destruction of the city of St. Pierre and the
death of 28,000 inhabitants; Mount Lamington (Papua New Guinea)
in 1951; Bezymianny (Kamchatka) in 1956; Arenal (Costa Rica)
in 1968; Mount St. Helens (Washington) with numerous explosive
eruptions in the 1980s; El Chichón (Mexico) in 1982; Nevado del Ruiz

VOLCANIC DANGER AND VOLCANIC RISK

Volcanic danger refers to the probability that a certain area around a certain volcano is subject to
a certain phenomena. Volcanic risk is a far broader concept, including an evaluation of the consequences of an eruption on the inhabitants and the environment. Volcanic risk can be defined using
the following formula:

$$\text{Risk} = \text{danger} \times \text{value} \times \text{vulnerability}$$

in which value refers to that which potentially could be lost in terms of human life, structures and
environmental factors, while vulnerability refers to the amount of the value believed to be in danger
of loss. In other words, the definition of volcanic risk is a synthesis of what is known about the volcano, the level of urbanization in the area, the state of the infrastructures, even the psychological
readiness of the people. For these reasons, mitigating the volcanic risk to an inhabited area requires
not just scientists armed with detailed knowledge of the volcano, but awareness of the dangers
related to an eruption on the part of the local population, the implementation of monitoring devices
and the development of emergency plans by the authorities.

(Colombia) in 1985, with the destruction of the town of Armero and
the death of more than 22,000 inhabitants; Pinatubo (Philippines) in
1991; Rabaul (Papua New Guinea) in 1994; Soufrière Hills (Montserrat, West Indies) in 1995–97. To this must be added the volcanic lake
of Nyos in Cameroon in 1986 which released a cloud of toxic gas that
killed 1,700 people. The total number of humans killed during the
last century amounts to about 75,000 (see table on page 84); the current century has already seen notable eruptions: Bromo in Indonesia
in 2000, Etna in 2001 and Nyiragongo in the Congo in 2002.

VOLCANIC RISK

Many of the world's active volcanoes are located in remote regions
that are inhabited by relatively few people. Such is the case for
most of the active volcanoes in Kamchatka and Alaska, which are
located in relatively uninhabited areas. There are many instances,
however, of active volcanoes located near settled areas. In many of

these situations, a volcanic eruption poses a great hazard to the inhabitants.

Large cities subject to volcanic risk include Seattle, in Washington State, one of the most populous cities of the United States, at risk from the possibilities of lahars from Mount Rainier; Quito, capital of Ecuador, threatened by the volcanoes Guagua Pichincha, Pululagua and Cotopaxi; Managua, capital of Nicaragua; San José, capital of Costa Rica; Manila, capital of the Philippines; Kagoshima, one of the most populous cities in Japan, located at the foot of the Sakura-Jima volcano. Central and southern Italy merit a discussion of their own, their active volcanoes putting them among the regions of the world most at risk. In Campania, Campi Flegrei and Vesuvius stand on the western and eastern borders of the city of Naples, where in some places there are population densities equal to those of Hong Kong. Campi Flegrei and Vesuvius are in a state of quiescence (for more than four and a half centuries for the first, only since 1944 for the second), a state that scientists interpret as a phase of magma reloading. In both cases the possibility of an explosive-type eruption is high, and such an event would inevitably involve a great number of people and would cause the destruction of vast urban and industrial areas. It is calculated that at least 600,000 people are at high risk in the case of an eruption from Vesuvius, with an equal number in danger in the area of Campi Flegrei.

On the island of Sicily, Etna constitutes a constant danger to the city of Catania with its population of half a million inhabitants. The city was partially destroyed by an eruption in 1669 during which lava flows reached the sea. An eruption today of similar force would have disastrous consequences. It must be borne in mind that Etna's eruptive history includes several Plinian-type explosive eruptions with the very highest potential for destruction.

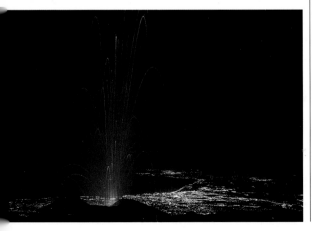

Left: Strombolian eruption at Etna. Visible in the background is the city of Catania.

MONITORING VOLCANOES

Today there are sophisticated instruments able to register many kinds of signals from volcanoes and provide information that helps scientists monitor volcanoes and predict their future activity. For several decades now, many of the countries threatened by volcanic activity have been making use of these instruments and perfecting them. In some cases volcanoes have been covered with a veritable network of instruments to collect various kinds of data that is then synthesized

CATASTROPHIC AND SIGNIFICANT ERUPTIONS

Year	Volcano	Country	Volcanic Activity	Other Activity	Victims
79	VESUVIUS	Italy	Ashfall, pyroclastic flows	Lahars	3,360
1586	KELUT	Indonesia		Lahars	10,000
1631	VESUVIUS	Italy	Pyroclastic flows	Lahars	3,500
1638	RAUNG	Indonesia	Ashfall, nuées ardentes		1,000
1672	MERAPI	Indonesia	Pyroclastic flows		3,000
1711	AWU	Indonesia		Lahars	3,000
1760	MAKIAN	Indonesia		Lahars	2,000
1772	PAPANDAJAN	Indonesia		Avalanches	2,957
1783	LAKI	Iceland	Ashfall, gas	Famine	10,521
1792	UNZEN	Japan	Gas	Avalanches, famine	15,188
1815	TAMBORA	Indonesia	Ashfall	Famine	92,000
1822	GALLUNGUNG	Indonesia	Pyroclastic flows		4,011
1856	AWU	Indonesia		Lahars	2,806
1883	KRAKATAU	Indonesia	Explosive eruption	Tidal waves	36,417
1888	BANDAISAN	Japan		Avalanches	461
1892	AWU	Indonesia		Lahars	1,532
1902	MT. PELÉE	Martinique	Pyroclastic flows		28,000
1902	SOUFRIÈRE	St. Vincent	Pyroclastic flows		1,680
1902	SANTA MARIA	Guatemala	Pyroclastic flows		6,000
1911	TAAL	Philippines	Hydromagmatic explosions		1,335
1919	KELUT	Indonesia		Lahars	5,110
1930	MERAPI	Indonesia	Pyroclastic flows		1,369
1938	RABAUL	New Guinea	Ashfall, hydromagmatic explosions		505
1951	LAMINGTON	New Guinea	Pyroclastic flows		2,942
1963	AGUNG	Indonesia		Lahars	1,900
1964	SURTSEY	Iceland	Phreatomagmatic explosive eruptions		
1973	HEIMAEY	Iceland	Explosive eruptions, lava fountains		
1980	ST. HELENS	U.S.A.	Explosive eruptions		60
1982	EL CHICON	Mexico	Pyroclastic flows, ashfall	Lahars	2,500
1985	NEVADO DEL RUIZ	Colombia	Explosive eruption	Lahars	22,000
1986	NYOS	Cameroon		Carbon dioxide gas	1,700
1991	PINATUBO	Philippines	Explosive eruption	Lahars	400

in a sort of radiograph. The best known example of this is the Hawaiian Volcano Observatory, where for many years now the level of monitoring has been able to predict a future eruption with a high level of accuracy. In Italy, the observatory at Vesuvius and the International Institute of Volcanology in Catania have among the most modern monitoring devices in the world, enabling them to predict the reawakening of the volcanoes of southern Italy with several days' warning.

Thanks to such monitoring, the civil authorities of many nations have been able to avoid major catastrophes. Such was the case of the 1991 eruption of Pinatubo (Philippines), where the type of event and the areas at risk were predicted by volcanologists well in advance, making possible the evacuation of the endangered population. Other examples of catastrophes averted include the 1994 eruption of Rabaul (Papua New Guinea) and the 1995 eruption of Soufrière Hills on the island of Montserrat (West Indies). The extreme case is the Sakura-Jima volcano (Japan), located only 5 miles (8 km) from the city of Kagoshima, where the local radio broadcasts news of the state of the volcano along with the regular weather report, keeping the population advised of imminent eruptions and alerting those areas of the city that will be subject to volcanic ashfall.

METHODS OF VOLCANO MONITORING

To be effective, volcano monitoring must be able to answer three different questions: 1) Where will the eruption take place? 2) When will it happen? 3) How large will it be?

Responding to these questions requires a deep understanding of the volcano in question, its eruptive history, its current state and its behavior before, during and after previous eruptions. In this sense, volcano monitoring unites all of the various fields of study that are part of the science of volcanology.

Much of the scientific research done by volcanic observers is directed at recognizing the sequence of events that precede an eruption. Such events are called volcanic precursors. In fact, for an eruption to occur, magma must rise within the crust of the Earth and also within the volcanic structure itself. The rocks that constitute the volcano and the area around it must in some way adapt to the arrival of the magma. The movements that result from this can be registered and measured. Furthermore, as it rises magma loses some of the gases it initially contains, and does so in proportions that differ according to its depth and chemical composition. The discharged gas reaches the surface following a more or less complex upward movement that involves interaction with the rock it passes. This gas, too, is sampled and analyzed. Finally, as new magma accumulates in upper areas it constitutes a disturbance to magnetic and gravitational fields in the vicinity of the volcano, adding further signs that can be recognized and interpreted.

Above: On Stromboli in the Aeolian Islands a surveillance camera has been installed on the peak of the volcano. The camera transmits "real-time" images of the volcano's activity 24 hours a day to a monitoring center in the city of Catania, where the pictures are closely analyzed.

In addition to these techniques there is the visual monitoring of the volcano, carried out during periodic excursions by volcanologists or made using a network of cameras that send real-time images of the volcano to the observatory. Such is the case of Etna on Sicily and Stromboli and Vulcano in the Aeolian archipelago.

VOLCANIC SEISMIC ACTIVITY

The period before an eruption is often marked by a large number of earthquakes (seismic clusters); these are produced by the interaction between the rising magma and the rocks through which it is moving. Studying the arrival times of the seismic waves and the characteristics of the seismic signals registered in various stations that form the monitoring network, scientists can determine several fundamental parameters of the volcanic activity, including its geographic location (epicenter) and its underground depth (hypocenter). It has been observed during many volcanic eruptions that the earthquakes perform an upward migration, which has been interpreted as a result of the progressive fracturing of rock toward the surface. By studying the seismic signals scientists can determine the depth of the magma and the speed at which it is rising inside the volcanic structure. Finally, using a mathematical model of the seismic signals it is possible to obtain information on the mechanical and geometric characteristics of the volcanic system in depth, information that is fundamental for predicting the characteristics of a future eruption.

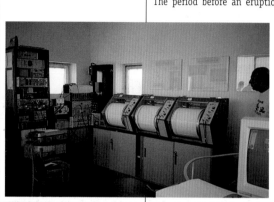

Above: Seismic analysis center in the Vesuvian observatory in Naples, Italy. Transmission stations that collect seismic data have been set up on Vesuvius, Campi Flegrei and the island of Ischia, and the collected information is studied here.

Because of the enormous value of seismic data, scientists do not limit themselves to measuring signals caused by earthquakes and instead produce waves of their own, setting off specially located explosive charges to create seismic waves that cross the volcanic structure and can be recorded on seismographs and analyzed in every detail. Of particular importance and great interest is the seismic tomography made at Vesuvius in the 1990s. This project involved setting off a high number of explosions; the seismic waves that were created were then measured by a large number of seismographs located on the volcano and in the surrounding area. In this way the scientists hope to eventually create a sort of three-dimensional image of the mechanical characteristics of the rocks inside and beneath Vesuvius. In theory, scientists could use such knowledge to piece together the primary characteristics of the volcano and use that information to localize and evaluate the size of any future accumulation of magma.

SURFACE DEFORMATION

As magma rises toward the surface it causes the swelling and deformation of the volcanic structure. The size and shape of such a deformation depend on the mass of magma in movement and the depth at which it is located. Instrument measurement of the deformation thus constitutes an effective method of following the approach of magma to the surface before an eruption. Many techniques are used to monitor surface deformation, including benchmarks and reflector sites, geodetic leveling, tiltmeters and global positioning system (GPS) measurements. The first two are used to periodically record and measure any variations in height or horizontal movement of the surface of a volcano. Tiltmeters can register even the variation in the inclination of slope of a volcano. The GPS system uses geo-

stationary satellites to establish the height and coordinates of recording sites and does so with remarkable precision. As monitoring tools capable of helping to predict eruptions, tiltmeters and GPS receivers have the advantage of providing a constant flow of data that can be used to determine the depth of magma in a volcanic system. In addition to the measurements that can be made in the field, several satellite technologies have recently been developed that use comparison of radar images taken at different times (Satellite Radar Interferometry and Synthetic Aperture Radar) to make precise measurements of surface deformation.

Above: Equipment for sampling volcanic gas set up in a fumarolic area on the island of Vulcano, Italy.

GEOCHEMICAL ANALYSIS OF FLUIDS

The gas discharged by magma in the crust of the Earth provides a great deal of information concerning the type of magma and its movement as well as information about the chemical, thermal and mechanical state of the rocks it is moving through. Monitoring volcanic gas is therefore of fundamental importance to assessing the state of a volcano and the magma inside it. This monitoring is carried out through the continuous or periodic collecting of samples of various fluids that contain a percent of the magmatic gas. Such fluids include the gas discharged from the crater of a volcano or from fumaroles, as well as gas emanations from the surface or from springs. The qualities that are measured include the chemical composition of the fluid, its temperature, the amount discharged over time, as well as numerous other measures of a more complex nature, such as the state of oxidation and the isotopic composition.

The approach of a volcanic eruption is usually preceded by an increase in the total quantity of gas discharged and by variations in the chemical composition of the gas, with an increase in its magmatic properties.

DETERMINING VOLCANIC DANGER

Knowledge of what to expect from any given eruption is essential to mitigating the risk. In reality, the impact of a volcanic eruption on the surrounding territory changes enormously from eruption to eruption. The preceding chapters have described the kinds of eruptions, both effusive and explosive, the great variability in volcanic processes, and the broad spectrum of possible types of eruption that can occur with different volcanoes, with different eruptions from the same volcano, and even with different stages in the same eruption. This information should make it clear how important it is to know the kind of eruption that is expected in order to evaluate and prepare for the associated risk. To do so requires detailed knowledge of the past history of the volcano and its current state, and the ability to make predictions of its future behavior. The monitoring activities described thus far are designed most of all to answer the questions of where and when a future eruption will happen. Knowledge of a given volcano's eruptive history combined with a physical-mathematical computer simulation of the eruptions serve to answer the question of how the eruption will take place.

The history of a volcano can be reconstructed on the basis of the type and extension of volcanic deposits. That information makes it possible to identify the areas that have been most frequently subject to certain eruptive processes and to establish for each area criteria of danger based on the frequency and intensity of the volcanic activity. Furthermore, knowledge of a volcano's eruptive history

Opposite: Computerized models of eruptions make it possible to predict which areas of a region will be most affected by certain kinds of volcanic phenomena. This figure combines a satellite image of the area of Vesuvius with superimposed lines indicating the areas of the surrounding landscape where the accumulated ashfall would be more than enough to cause the collapse of the roofs of homes.

makes possible the evaluation of its current state and prediction of various important parameters of any future eruption. For example, knowing the composition of the magma is essential to any evaluation of future volcanic processes. On the basis of a model of the current state of the volcano, based on the assembly on stratigraphic, chemical, physical and structural knowledge of the volcano and its products, the computer simulation of the eruption permits study of the dynamics of the expected eruption and its impact on the surrounding environment. The volcanic system and the processes that characterize the rise of magma and its arrival on the surface are represented using the laws of physics and chemistry, using sophisticated computer programs that use equations to describe such processes. In this

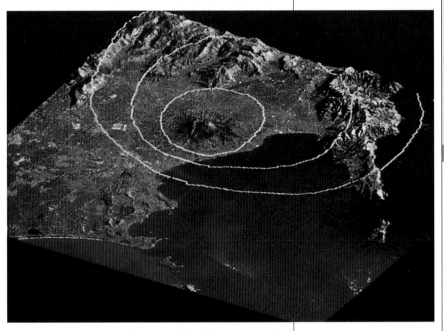

way it is possible reach a prediction of the anticipated eruptive processes, of the different areas that each will effect, and of the time period connected to each dangerous event. Furthermore, the model makes it possible to work up a large number of physical measurements that will characterize the expected eruption and the associated phenomena and thus to reach an estimate of the danger posed. This estimate is then put together with the estimate made on the basis of stratigraphic studies to create a final synthesis of all that is known about the volcano in question.

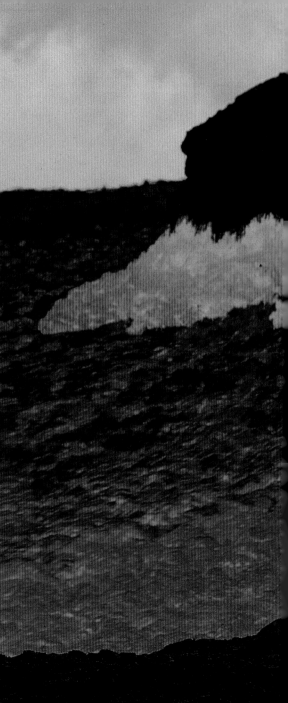

THE VOLCANOLOGIST

Preceding pages: A volcanologist in a heat-resistant suit on Kilauea, Hawaii.
Above: Positioning a displacement meter to measure changes in a fracture at the crater of Fossa, on Vulcano, Italy.
Right: Taking a sample from a lava channel on Etna.

The study of volcanoes is a fascinating adventure. An active volcano is an open door to the inside of the Earth, to the primordial forces that shaped our planet and are determining its evolution; it is a way of understanding the world in which we live, the mechanisms that made it what it is today and that are constantly acting to change it. An active volcano is among the most complex physical systems in nature. Volcanic processes take place on a time scale that runs from millennia to fractions of a second, with a spatial scale that goes from hundreds of miles to fractions of an inch.

The technologies used to study active volcanoes involve a vast gamut of scientific disciplines, among them geology, petrology, geophysics, geochemistry, hydrology, fluid mechanics, thermodynamics, engineering, meteorology and information systems. More and more often the important developments in volcanology come from the combination of two or more disciplines. For this reason, defining the figure of today's volcanologist is not a simple matter; in reality, a conference on volcanology is a gathering of people from the most disparate scientific fields, scientists who carry out their professions using extremely different methods, but who are all dedicated to achieving the same goal: improved understanding of volcanic systems.

THE "CLASSIC" VOLCANOLOGIST

Historically, the study of volcanoes came into being from the need to interpret volcanic rocks and the geological forms related to the presence of a volcano. Therefore it was, and still is, a matter for geologists, who collect, sample, analyze and interpret volcanic rocks, as well as map volcanic areas and structures. In time this role became more specialized,

Left: Sampling lava during an effusive eruption in 1985 on Stromboli (Aeolian Islands, Italy).
Below: Volcanologist Mauro Rosi of the University of Pisa, one of the authors of this book, on Stromboli, a volcano that offers scientists the opportunity to make close-up observations of many eruptive phenomena.

and the field of the volcanologist evolved into a kind of scientist-detective who examined every tiny residue and every smallest detail contained in volcanic deposits to reconstruct eruptive events. Today the role of the geologist who studies volcanoes has become even more involved and complex. The existence of computerized models of volcanic eruptions requires a large amount of precise data concerning specific eruptions and also demands continuous "real-time" checking on the actual terrain to determine the accuracy of predictions and their effective application to reality. As a consequence, the volcanologist has become more and more a scientist who studies the physical processes that determine and guide an eruption and that lead to the production of volcanic deposits of certain types. The day-to-day equipment of the classic volcanologist has remained unchanged: the rock hammer, shovel and compass are still his or her constant companions; but the necessary scientific training embraces increasingly vast areas, and the deposits that the volcanologist is called upon to "read" speak more and more in the language of physics.

THE GEOCHEMIST

The geochemical study of magma originally came into being as a branch of petrology to study the chemical behavior of the trace elements (those usually present in only very small quantities) found in magma. There is in fact a correlation between the relative abundance of such elements, the overall chemical composition of the

magma, and the geodynamic environment in which the magma was produced. More recently, geochemistry has turned its attention to other components of magma, also usually present in small quantities: the volatile elements. The geochemist is thus a geologist or chemist willing to occasionally risk his or her life—and to daily risk his or her health—during hours spent taking samples of highly noxious volcanic gas inside fumaroles at temperatures of hundreds of degrees to then bring the resulting collection of vials of poisonous liquids back to a laboratory where their contents can be analyzed with extreme care to avoid atmospheric contamination.

The geochemist's work is of enormous importance. In fact, information drawn from the study of volcanic gas can be used to evaluate the state of a quiescent volcano and potentially to predict its future behavior. The rise of magma from the depths of the Earth toward the surface, a process that obviously precedes an eruption, causes changes in the physical state of the magma that are reflected in the magma's chemical balance and that thus show up in the composition of the gas the magma emits.

Above: Sampling volcanic gas. The geochemist wears a protective gas mask.
Right: Geochemists on the island of Vulcano, Italy.

Other activities performed by the geochemist include measuring the total quantity of gas emitted by a volcano into the atmosphere, both while quiescent and during an eruption. Among the instruments used to achieve this are spectrometers installed on satellites and airplanes.

THE GEOPHYSICIST

Within the field of volcanology, the geophysicist makes use of several disciplines, including gravimetry, magnetometry, geoelectrics and most of all seismology. The great importance of seismology is based on the fact that movements of magma deep in the Earth cause deformations of the rocks above and around the magma; such movements create seismic waves that can be recorded, measured, analyzed and interpreted to determine the depth of the magma and the extent of its movement. This makes possible information on the eventual rise of the magma and also on its current position, on the existence and size of a future magma chamber, and on the mechanical characteristics of the rocks beneath a volcano—indispensable data for defining a volcanic system. An outstanding example of this process is the reconstruction of the system of alimentation of the Kilauea volcano in the Hawaiian archipelago, most of which was based on the analysis of seismic signals recorded on the island.

THE EXPERIMENTAL VOLCANOLOGIST

The role of the experimental volcanologist is to reproduce in a laboratory (on a reduced scale) various important volcanic processes, such as those involved in the movement of magma, the emission of gas from magma, the dispersal of pyroclasts into the atmosphere, the collapse of a caldera, and so on. The aim is to simulate a natural process on a scale and under conditions that permit its close continuous observation so as to derive information on what happens and how it happens. Given the small scale, precise mathematical relationships can be established for the various quantities involved, and this detailed information can then be applied to the natural system itself.

Carrying out such laboratory experiments is of great importance, but the problems to be resolved are often beyond the scope of models. In fact, many of the processes described in this book simply cannot be reproduced on a reduced scale, at least not with the technology available to us today. A good example is the geometry of a volcanic conduit, many miles in length but only a few dozen yards wide, at the most perhaps a few hundred. There is the fact that most important volcanic processes take place over such long periods of time that they could never accurately be reproduced. And the energy that is given off during a volcanic eruption reproduced on even a very small scale would

Above: Deviation of a lava flow on Etna in 1992 by volcanologists from the Gruppo Nazionale per la Vulcanologia, led by Franco Barberi, a famous international volcanologist and at the time director of Italian civil defense. History's first successful deviation of a lava flow, the effort was made to redirect the advance of a lava flow that threatened the town of Zafferana Etnea.

easily destroy more than an entire laboratory. Even so, a great deal of invaluable information has been obtained in the experimental field. For example, much of what we know today about the modalities of the deformation and flow of lava came from experiments conducted in modern laboratories; the same is true of our knowledge of how volcanic gas is released. At the present time experiments are being conducted to simulate the process of the fragmentation of magma inside a volcanic conduit and the explosive interaction of magma and water.

Right: Use of a portable seismograph. Below: A young volcanologist examining a hornito of carbonatite lava at Ol Doinyo Lengai in Tanzania.

THE COMPUTER ANALYST

The role of the computer analyst is that of creating suitable systems and models of equations to describe the processes that occur before, during and after an eruption and to employ such equations to determine the distribution of the physical parameters and therefore the dynamics of volcanic processes. Many scientific disciplines are applied to this work, including fluid dynamics, thermodynamics, chemical kinetics, structural mechanics, the rheology of magma (the study of its deformation and flow) and so on. In addition to this, knowledge of information systems is necessary, since the physical models capable of representing such complex systems as volcanoes make use of mathematical formulas that require the use of computers.

The computer analyst is clearly the newest addition to the many volcanology experts discussed here, and the analyst's role has increased proportionally as computers have become faster and faster, operating with more and more memory. The computer analyst is very often a physicist or engineer, sometimes a mathematician, more rarely a geologist with the necessary training in physics. The problems for the computer analyst to examine and solve include the rise of magma from the mantle toward the crust, the formation of a magma chamber, the physical and chemical processes that take place within the chamber, the fracturing of the rocks that opens the way for the further rise of

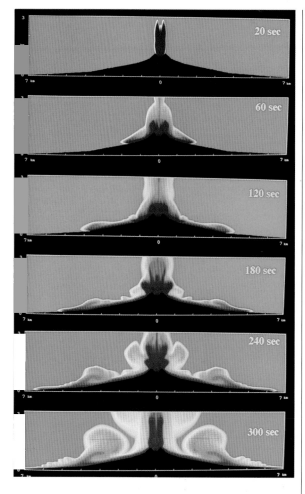

60 sec

120 sec

180 sec

240 sec

300 sec

Left: Computer simulation of an eruptive event with characteristics based on those believed possible in the event of a future eruption of Vesuvius. The images are arranged in a time sequence beginning near the onset of the eruption and indicate the concentration and dispersal into the atmosphere of volcanic materials. The pyroclastic material can be seen to rise upward at high speed. In this case, however, the eruptive conditions used in the simulation were such to produce the gravitational collapse of the eruption column with the consequent formation of a pyroclastic surge that moves off the slopes of the volcano at high speed. At a certain distance from the crater volcanic gas begins to separate out of the pyroclastic surge and rise, dragging some of the finer material upward to generate the beginning of a co-ignimbrite cloud.

Below: The volcanic area of Lake Myvatn, Iceland. Visible in the immediate foreground are illustrations with notes on volcanic and hydrothermal activity put in place by Icelandic volcanologists for tourists and other visitors to the site.

the magma, the way it rises, the fragmentation of magma, the dynamics of an eruption, the dynamics involved in the release of gas, the explosive interaction of magma with underground or surface water, the rise of a Plinian column or its collapse, the dynamics of pyroclastic flows and lava flows and the evolution of a lava field.

All of these are questions that are in fact being answered as analysts prove themselves capable of gathering and deciphering vast quantities of information on the dynamics of volcanic processes, information that could not have been obtained in any other way and that is making a decisive contribution to our understanding of volcanoes and volcanic eruptions.

SYMBOLS

The following pages present descriptions of the 100 most famous and most representative volcanoes. Each entry gives the volcano's name, its geographic location, its latitude and longitude, its altitude (based on the highest area of the volcanic region) and its reference number. This number is composed of two series of numbers, separated by a hyphen, the first identifying the volcano's geographic location, the second giving its position among the volcanoes of that area.

The geographic information, history of eruptions of the volcano and reference numbers are based on *Volcanoes of the World*, published in 1994 by the Smithsonian Institution in Washington, D.C., integrating that information with more recent data from the Smithsonian, including scientific publications and bulletins on the state of activity of the volcanoes.

Each volcano is also classified on the basis of two fundamental criteria, its geodynamic environment and the type of volcanic structure. This information is given with the use of symbolic diagrams (see pages 100–101).

The description of each volcano includes the volcano's prevalent types of activity and such general information as its structure, geological evolution and eruptive history, including principal eruptions and the effects on nearby inhabited areas.

Also given is the information necessary for visiting the actual site, including indications of the area's main attractions.

The information provided for visiting the volcanoes includes the easiest way to reach the area of the volcano and the major reasons for going; also indicated is the level of difficulty involved in such a visit.

It cannot be overemphasized that the relative safety in climbing an active volcano is directly related to the state of activity of the volcano, and such states can change enormously over time. It must also be remembered that volcanoes are mountains that present particular dangers, such as concentrations of gas, thermal areas, avalanches, falling rocks, etc.

Therefore it is highly advised that any trip be carefully planned and based on information from specialized guides and personnel in the area itself, including reference to the most up-to-date information on the state of activity of the volcano obtained from the people or organizations actively monitoring the volcano, such as observatories and research organizations, a list of which is given on pages 328–31.

GEODYNAMIC SETTING

Each of the volcanoes listed here is found in one of the five geodynamic settings presented in the diagram below. There are a few volcanoes, primarily on the Italian peninsula (Campi Flegrei, Vesuvius, Etna), for which a geodynamic setting is not given because the situation is too complex to be easily associated with one of these general categories. A more detailed description of each setting can be found in the introduction.

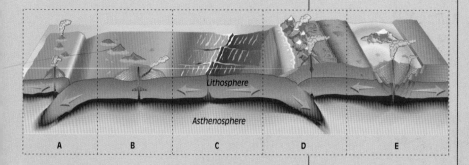

A. **ISLAND ARCS** (e.g., Aleutians, Japan, Indonesia) When two oceanic plates meet, one of them is subducted beneath the other, being drawn into the mantle and generating masses of magma that rise upward to form island-arc volcanoes.

B. **HOT SPOT** (e.g., Hawaii, Galápagos, Azores, Yellowstone) An area in the middle of a plate where magma rises from the mantle; the magma erupts through the plate, creating a series of volcanoes whose alignment indicates the movement of the plate over the hot spot.

C. **OCEANIC RIDGE** (e.g., Iceland) Magma rises along a broad system of oceanic fissures and generates volcanic structures, which in some cases rise out of the sea.

D. **CONTINENTAL MARGIN** (e.g., Cascades, Andes) In the collision of an oceanic plate with a continental plate, the oceanic plate is subducted beneath the continental plate. As it is subducted into the mantle the oceanic plate sets off the production and rise of magma, leading to the formation of a chain of coastal mountains, including volcanoes.

E. **CONTINENTAL RIFT** (e.g., East African Rift Valley) As tectonic plates separate a new ocean is formed by the thinning of the crust; magma rises along the system of fissures.

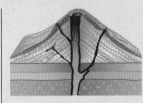

Stratovolcano

Compound volcano

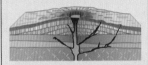

Shield volcano

Caldera

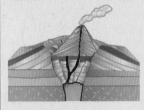

Somma volcano

Fissure volcano

VOLCANIC STRUCTURES

The volcanoes listed here are divided into 10 types according to their structure: stratovolcano, compound volcano, shield volcano, caldera, somma volcano, fissure volcano, lava dome, cinder cone, tuff cone and table mountain.

The icons that identify the type of volcanic structure appear along with those indicating geodynamic setting. The chapter on volcanic structures provides more information on the major types of volcanoes.

STRATOVOLCANO
Volcanic cone with steep sides built up from layers of material from lava and pyroclastic flows, most of it erupted from a central vent, with additional layers of material from mudflows and lahars. Also called composite volcano.

COMPOUND VOLCANO
Volcanic structure often of large size and consisting of two or more vents formed by the association of such diverse volcanic structures as lava domes and stratovolcanoes often formed at different times.

SHIELD VOLCANO
Large structure with gentle slopes, giving it the shape of a warrior's shield; formed almost entirely of layers of lava flows and often crossed by eruptive fissures along which craters and cinder cones align.

CALDERA
Large circular depression shaped like a cauldron or basin produced by the collapse of a volcanic structure, usually the roof of a magma chamber.

SOMMA VOLCANO
Volcano with a structure similar to that of the Somma–Vesuvius complex, composed of a large summit depression surrounded by steep scarps inside which a new volcanic cone has been formed.

FISSURE VOLCANO
Structure composed of the accumulation of volcanic material on the two sides of an eruptive fissure, usually found in regions of diverging plates (or on an oceanic ridge).

LAVA DOME
Mass of lava with steep sides produced by the accumulation of viscous lava around an eruptive vent.

CINDER CONE
Volcanic cone formed from the accumulation of scoria emitted during Strombolian activity.

TUFF CONE
Volcanic cone composed of the accumulation of ash, scoria and pumice emitted by explosive activity of the hydromagmatic type.

TABLE MOUNTAIN
Volcanic massif with steep sides and horizontal summit created following an eruption of lava under a glacial covering.

PREVALENT VOLCANIC ACTIVITY

In terms of activity and related phenomena most volcanoes display a variety of behaviors. Much volcanic activity is somewhat uniform in the sense that it involves repetitions of eruptions with very similar characteristics. Volcanoes like Stromboli and Etna or those of the Hawaiian Islands are known for eruptions of lava accompanied by low-intensity explosive activity, while others, such as Campi Flegrei, are thought to have been produced almost exclusively by explosive eruptions. All the same, over the long course of their existence the majority of volcanoes have demonstrated eruptive activity along with very different phenomena, ranging from peaceful emissions of lava flows to gigantic explosive eruptions to hydrothermal phenomena, the last being a highly variable type of activity.

The prevalent activity given for each volcano has been established on the basis of its most recent and most frequent eruptions.

In each entry this activity, divided into four basic categories, is given in accordance with the following definitions. The chapter on volcanic eruptions has more detailed information on the more important of these activities.

EXPLOSIVE ACTIVITY
Hawaiian, Strombolian, Vulcanian, Pelean, hydromagmatic, phreatomagmatic, Plinian, pyroclastic flows.

EFFUSIVE ACTIVITY
Lava flows, lava lakes, lava domes.

HYDROTHERMAL ACTIVITY
Fumaroles, phreatic explosions, hot springs, geysers.

RELATED PHENOMENA
Lahars (mudflows), avalanches, caldera collapses, tsunamis (tidal waves), acid rain, gas clouds.

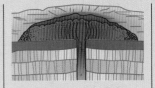

Lava dome

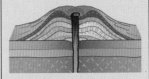

Cinder cone

Tuff cone

Table mountain

Most of the active volcanoes in Europe are in one of three broad areas: the Atlantic, the Mediterranean and Turkey, also including the mountains of the Caucasus. There are a few Holocene volcanoes (meaning relatively recent, less than 10,000 years old) in the Massif Central of France (Chaine Des Puys) and in the valley of the Rhine (West Eifel Volcano Field) in Germany.

Active volcanoes in Turkey and the region of the Caucasus

The volcanoes in this zone, about 20 all together, are stratovolcanoes and basaltic cinder cones produced by single eruptions. Only two stratovolcanoes (Nemrut Dagi in Turkey and Elbrus in Russia) have erupted within the last 2,000 years.

Active volcanoes in the Atlantic area

The active volcanoes in the Atlantic are in Iceland, the Canary Islands (Spain), the Azores and the Cape Verde Islands (Portugal). Iceland is a volcanic island located along the Mid-Atlantic Ridge with a high concentration of active volcanoes (24 have erupted during the past hundred years). Much of the activity consists of dramatic emissions of basaltic lava from open fissures, results of the ongoing separation of the North American and Eurasian plates. There also have been explosive eruptions with emissions of pumice, such as that of Askja in 1875. One of the features of the volcanic activity in Iceland is subglacial eruptions; those that take place under the large Vatnajökull glacier melt ice and cause large-scale flooding (*jökulhlaup*).

The lifting of the Mid-Atlantic Ridge above the surface of the water to form Iceland results from the presence of a large hot spot under the ridge.

The island groups of the Canaries, Azores and Cape Verdes are also located above hot spots. The hot spot of the Azores is on the Mid-Atlantic Ridge at the spot where three plates—the North American to the west, Eurasian to the northeast and African to the south—are in contact, while the hot spots that formed the Canaries and the Cape Verdes are to the east of the Mid-Atlantic Ridge inside the African plate. The volcanic activity in all the archipelagos has produced large central shield volcanoes created by the repetitive overlayering of flows of basaltic lava. In their later stages of development some of these have erupted viscous magma rich in gas, with occasional explosive eruptions, and some such as Pico de Teide on the island of Tenerife in the Canaries have large summit calderas.

During the phase when a volcanic island is rising out of the water, the eruptive activity is often dominated by phenomena caused by the interaction of seawater and magma (hydromagmatic explosive eruptions).

Taken all together these three archipelagos have about 15 volcanoes that have erupted in historical times; about 10 of those are in the Azores.

Subglacial Arctic Ocean

ICELAND

20. Herdubreid
19. Krafla
18. Askja
17. Laki
16. Grimsvötn
15. Hekla
14. Heimaey
13. Surtsey
12. Snaefellsjökull

EUROPE

1. Campi Flegrei
2. Somma Vesuvio
3. Stromboli
4. Vulcano
5. Etna
6. Santorini
7. Nisyros

CAUCASUS

AZORES
8. Pico
9. Agua De Pau

Tenerife
Hierro

CANARIES

TURKEY

MEDITERRANEAN SEA

RED SEA

Tropic of Cancer

AFRICA

The active volcanoes in the Mediterranean area

About 20 volcanoes are located in the Mediterranean area, most of them on the Italian peninsula and Sicily, in the sea around Sicily, and in the Aegean Sea. The volcanism of the Aeolian Islands and the islands in the Aegean results from the ongoing processes of approach and collision between the African and Eurasian plates; also involved is the subduction of a portion of the African plate under the Eurasian, with the production of magma typical of island arcs. The volcanoes of the Aegean Sea and the Aeolian Islands share the typical arcuate structure and enclose a small area of seawater.

Etna, the most active volcano in Europe, incessantly erupts basaltic lava along with lava of an intermediate composition of the sodium-alkaline type. Etna's location on the Ionian side of Sicily is believed to result from the process of subduction and fracturing of the eastern edge of the island against the basin of the Ionian Sea. The volcanoes in the area of Naples (Vesuvius, Campi Flegrei and Ischia) erupt alkaline-potassium lava. Many scientists believe this is related to a late stage in the process of collision with the immersion of the subducted plate from east to west, associated with the fracturing of the border of the Italian peninsula produced fol-

lowing the opening of the Tyrrhenian Sea, with the consequent counterclockwise rotation of the peninsula.

Because of their location in an area that has been inhabited since antiquity, the volcanoes of the Mediterranean area have long histories of interactions with humans. The cataclysmic eruption of Santorini (ancient Thera) in 1650 B.C., which buried the city of Akroteri, and that of Vesuvius in A.D. 79, which buried the cities of Pompeii and Herculaneum, are universally famous examples of volcanic disasters. Both have also provided archaeologists with a wealth of information documenting the life and organization of those important early civilizations.

The Italian volcanoes (Vesuvius, Stromboli, Vulcano, Etna) have attracted generations of scientists who, with their studies, laid the basis for the science of volcanology. It was on Vesuvius, in 1841, that the world's first volcanological observatory was created.

CAMPI FLEGREI, ITALY
Lat. 40.83 N – Long. 14.14 E – Alt. 1,503 feet (458 m) – 0101-01

Caldera

Prevalent volcanic activity. Plinian, phreatomagmatic, Strombolian, lava domes, fumaroles.

General. The area of Campi Flegrei is the site of a high number of volcanic cones and craters. This is a caldera 7½ by 9½ miles (12 by 15 km) in size created following two gigantic explosive eruptions, one 36,000 years ago and the other 14,000 years ago. The slopes of the original depression are still visible from Monte di Procida along the northern side of the Piano di Quarto and at the Collina dei Camaldoli. To the south, the margins of the caldera are lost in the Gulf of Pozzuoli.

The most ancient of the two large eruptions, named for the Campanian ignimbrite it erupted, deposited between 24 and 36 cubic miles (100 and 150 km³) of magma of trachytic composition. The second eruption, named for the Neapolitan yellow tuff it erupted, emitted a volume of magma equal to 4¾ to 7¼ cubic miles (20 to 30 km³). Over the course of the past 10,000 years the eruptive history of Campi Flegrei has been marked by frenetic periods of activity separated by periods of rest lasting centuries and millennia. The last large eruption took place between 4,500 and 3,700 years ago: numerous vents opened to the northeast of Pozzuoli and near the present-day site of Lake Avernus. The activity was predominantly explosive, with a significant level of interaction with water, although there were also episodes of lava emission (domes). The most recent eruption, which created Monte Nuovo ("new mountain"), began at one o'clock in the morning on September 29, 1538, near the old town of Tripergole on the eastern banks of Lake Avernus, where no eruptive center had previously existed. Over the span of a few days the eruption created a mountain of ash and pumice that stands 427 feet (130 m) high.

The name Campi Flegrei ("burning fields") refers not so much to volcanic activity (after all, during the classical age the area had been dormant for 1,500 years) as to the presence of many fumaroles and hot springs, areas of steam emission and boiling mud, the results of a very hot underground rich with liquids. The ancient Romans made much use of the hot springs.

As a result of bradyseism (the slow uplifting and sinking of ground) many of the structures dating back to Roman times are today below sea level. In the period 1982–84 the city of Pozzuoli experienced uplifting. The causes of bradyseism are not yet fully understood, but volcanologists cannot rule out the possibility that the recent events of uplifting may be in some way related to increased pressure in the magma chamber beneath Campi Flegrei. Such phenomena would imply the danger of a return to activity. The urban expansion of the area over the course of the last century (more than half a million people now live inside the area of the caldera) has dramatically increased the potential dangers and the risk factors.

Access and principal attractions. Campi Flegrei offers numerous opportunities to see tuff rings, lava domes, pyroclastic deposits and thermal activity. The famous Solfatara crater is the site of spectacular hydrothermal phenomena, including fields of fumarolic emissions and boiling mud pots. There is a fine example of a tuff ring at the Astroni crater, one of the few areas still not contaminated by urban sprawl and formerly a hunting reserve of the king of Naples. Excellent examples of tuff cones produced by the interaction of water with magma include the cones of Monte Barbara, Capo Miseno and Nisida. The tuff ring of Bacoli and the crater of Avernus are enchanting natural settings, while one of the outstanding archeological sites of the region is to be found near the Cuma lava dome. The western slope of Monte Procida offers spectacular views of stratigraphic pyroclastic deposits, including banks of Plinian pumice, deposits of breccia, lava domes, and pyroclastic flows and surges from the period between 50,000 and 14,000 years ago.

105

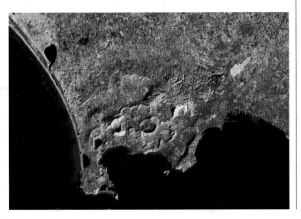

Opposite: The crater of Monte Nuovo, created in 1538. In the background is the city of Pozzuoli, located in the middle of the caldera.
Left: Satellite image of the area of Campi Flegrei. The numerous craters inside the caldera are clearly visible.

Compound volcano

Prevalent volcanic activity. Plinian, Strombolian, pyroclastic flows, lava flows and domes.

General. This stupendous volcanic complex, with the Bay of Naples as its dramatic background, represents in the collective imagination of the world the prototypical volcano as well as the prototypical example of volcanic destruction. Vesuvius, well known as one of our planet's thousands of active volcanoes, has over the centuries drawn to it generation after generation of chroniclers, scientists, archaeologists and artists. The very science of volcanology had its baptism on Vesuvius, for it is here that the world's first volcanological observatory was established, in 1841.

Beginning with the two famous letters written by Pliny the Younger, which stand as the first written reports of a volcanic eruption, Vesuvius has provided volcanology with much of its technical terminology, still in use, which further testifies to the major influence that the studies conducted on it have had on the literature of volcanoes.

The volcano is composed of two different volcanic structures: Monte Somma, an older, collapsed stratovolcano with an elliptical summit caldera 2.1 by 3 miles (3.4 by 4.9 km), and inside that caldera the cone of Vesuvius. Today Monte

Somma reaches an altitude of 3,707 feet (1,130 m), while the cone of Vesuvius is slightly higher (4,203 feet [1,281 m]). Information from geothermic soundings carried out on the southern slope of the cone indicate that the volcanic material is about 4,920 feet (1,500 m) thick. Based on dating of the oldest material found during the course of the same survey, the volcanic activity is believed to have begun about 300,000 years ago.

The Somma caldera is the result of a series of four Plinian eruptions that took place 18,000, 17,000, 8,000 and 3,360 years ago as well as, of course, the eruption of A.D. 79. This last event, which led to the destruction of the Roman cities of Pompeii, Herculaneum and Stabiae and in which Pliny the Elder lost his life, has become consecrated in history and stands as the best known and most thoroughly documented volcanic eruption of all time. It is also, of course, the prototypical Plinian eruption. According to the eyewitness reports in the two letters that Pliny the Younger sent the Roman historian Cornelius Tacitus, the event began on August 24 at 1 P.M. and ended about 8 P.M. of the next day.

Study of the deposits indicates that the eruption went through two distinct phases. The first, which lasted until 1 A.M., involved the formation of an enormous Plinian column, about 19 miles (30 km) high, that produced an incessant rain of pumice on the cities of Pompeii and Stabiae, causing the roofs of houses to collapse. The prevailing wind was to the southeast, carrying the material in the Plinian column (ash and pumice) directly toward Pompeii while sparing Herculaneum. Even today one can see deposits from this phase of the eruption accumulated in certain areas of the city. The second phase involved pyroclastic flows and surges that destroyed Herculaneum, burying it beneath more than 66 feet (20 m) of pumice and ash and destroying whatever remained of Pompeii.

The eruption of A.D. 79 was followed by innumerable Strombolian eruptions and effusions that led to the gradual construction of the Vesuvius cone and the deposits formed by lava flows on the southern and eastern slopes. Since then the eruptive activity has gone through only two important rest periods, both of which were followed by explosive events of enormous energy. The eruption of 472, which followed a period of rest that had lasted three centuries, was of the Plinian type and produced devastation on a scale comparable to that of A.D. 79. After another period of rest, which probably lasted until 1139, Vesuvius again became active, and at 7 A.M. on December 16, 1631, it

Access. A car is the easiest way to reach the crater since you can drive right up to the edge of the cone and find parking. There is then a path leading to the top. Along most of this path you walk on ash and lapilli left from the 1944 eruption. After about a 655-foot (200-m) climb you reach the edge of the crater, an enormous pit encircled by vertical walls. Visible on the opposite wall is the stratification left by innumerable lava flows from activity dating back to 1906. Above this are accumulations of pyroclastic material from the eruption of 1944. Moving around the edge of the crater in a counterclockwise direction gives you the chance to see the many fumaroles that discharge through fractures in the crater. The ideal way to descend is the circular route that takes you around the major cone (Gran Cono): go down toward Boscotrecase and keep left when you reach the height of a parking lot; this will lead you back to your point of departure. It is of course necessary at all times to obey the instructions of the guides, who have authority to control access to each of the many points of interest.

Opposite: Aerial view of the Gran Cono of Vesuvius and its summit crater. Behind that is the wall of the caldera of Monte Somma, with a view of the city of Naples in the background.

erupted. This was a Plinian eruption that continued until the afternoon of the following day. The area around the volcano was thoroughly devastated by raining lapilli and ash, pyroclastic flows and mudflows. The eruptive phenomena caused enormous damage and took the lives of about 4,000 people.

The volcano then returned to almost continuous Strombolian activity, permitting European scholars of the period to study the volcano close up. The great scientific and naturalistic attention attracted by the eruptions of Vesuvius over four centuries, together with the pressing need to protect the resident population from the danger of further eruptions, led in 1841 to the foundation by the king of Naples of the royal Vesuvian observatory, the world's first volcanological institute dedicated to the systematic observation and study of volcanic phenomena. Thus began the initial activity of monitor-

ing the volcano, destined to reveal itself as of fundamental importance to understanding the phenomena of volcanic eruptions and for predicting them.

After the most recent eruption, in March 1944, Vesuvius entered a state of repose that has now lasted more than half a century, although this rest cannot in any way be taken as an indication, much less as proof, of extinction. Unfortunately, despite the well-documented events of the past, the current situation, viewed from the point of view of protecting the population, is critically out of hand. The streets are clogged with uncontrolled new building, much of it illegal, permitting construction nearly up to the volcano's vent.

For this reason Vesuvius is considered by volcanologists and others to be one of the most dangerous volcanoes in the world, if not far and away the most dangerous.

Principal attractions. *Vesuvius, the most famous volcano in the world because of the eruption that destroyed Pompeii and because of its location in the splendid Bay of Naples, has always been a tourist mecca. The main attractions are the interesting lava fields located on the flanks of the volcano, which contain crystals of leucite; the pit of the crater produced by the eruption of 1944; and the Valle del Gigante covered by broom woods that grow up to the very base of the caldera's wall. The wall is cut by numerous vertical dikes.*

The climb along the crest of Monte Somma is particularly pleasant, offering splendid views of the Gran Cono and the lava flow generated by the 1944 eruption; the lava leads down the cone, expanding as it enters the Valle del Gigante.

109

Opposite: Satellite image of the Bay of Naples, showing its volcanic areas—Vesuvius at the center, Campi Flegrei to the right—the island of Ischia to the far left, and to the right the island of Capri and the Sorrento peninsula.
Left: Aerial view of the summit crater of Vesuvius, created during the last eruption, in 1944.

STROMBOLI, ITALY

Lat. 38.79 N – Long. 15.21 E – Alt. 3,038 feet (926 m) – 0101-04

Island arc

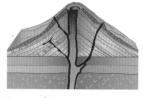

Stratovolcano

Prevalent volcanic activity. Strombolian, lava flows, landslides, fumaroles.

General. Although reaching only 3,038 feet (926 m) above sea level, the cone of Stromboli rises more than 9,800 feet (2,987 m) from the sea floor.

Stromboli has added its name to the vocabulary of science, Strombolian being the standard name for a certain category of explosive activity. It is a remarkable sight: from vents in the summit craters, located about 650 feet (198 m) below the peak, explosions occur at intervals of tens of minutes producing jets of incandescent material that are shot up hundreds of feet in the air. Most of this material falls back around the vents; another portion rolls down the steep slopes of the Sciara del Fuoco into the sea. Stromboli has been in this state of persistent activity for almost 2,000 years, a fact that is reflected in its nickname: Lighthouse of the Mediterranean.

The island of Stromboli was built up by volcanic activity, but initially from another volcano of which the present-day volcanic neck of Strombolicchio is another remnant; that volcano's eruptions ended about 200,000 years ago. Over time the activity moved a distance of about 1.85 miles (3 km) to the

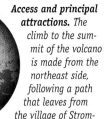

southwest, leading to the gradual construction of the current stratovolcano. The cone has undergone major changes during the last 13,000 years. Eruptive activity first caused the formation of the uppermost part of the island, the peak of Vancori; the important accumulations of lava on the northwestern slopes were produced later, in part a result of the opening of eruptive vents in the slopes of the cone (Timpone del Fuoco). During this same period at least three large-scale avalanches took place on the northeastern slope, dumping enormous masses of rocky debris into the sea. These avalanches, all of which took place in the same area, created the large depression today called the Sciara del Fuoco, inside which so much of the material erupted from the volcano accumulates.

The Sciara del Fuoco is bordered by steep rocky walls up to 985 feet (300 m) high, along which extend rock faces over hundreds of yards composed of radial dikes exposed by erosion. The depression of the Sciara del Fuoco continues into the sea to a depth of about 1,970 feet (600 m) below sea level, eventually becoming an underwater canyon.

Other than the normal Strombolian activity there are effusive eruptions and paroxysmal explosive eruptions. About every 10 years the volcano gives off lava flows that reach all the way to the sea; the last such occurrence was in 1985–86.

The paroxysmal eruptions are far more explosive, able to hurl blocks of rock and fragmented magma distances of several miles, even reaching inhabited areas along the coast; this happened during the eruption of 1930, the largest of Stromboli in the course of the past hundred years.

Access and principal attractions. The climb to the summit of the volcano is made from the northeast side, following a path that leaves from the village of Stromboli itself. The climb, easy up to an altitude of about 985 feet (300 m), follows a somewhat steep path that rises along the eastern edge of the Sciara del Fuoco. The descent is made following an easy trail that runs along the valley of the Rina Grande and ends near San Vincenzo. The primary reason for visiting Stromboli is the opportunity of witnessing the activity of the summit craters close-up. You can also climb to the upper parts of the volcano (Pizzo Sopra la Fossa), but this requires hiring an authorized guide; according to the amount of volcanic activity taking place, you may find visits to certain crater zones restricted. Among the other geological attractions is the Sciara del Fuoco, which reveals how avalanches create an amphitheater-shaped depression, and Strombolicchio, a fine example of a volcanic neck. The Civil Defense office in town offers more information in general about the volcano.

Opposite: Stromboli erupting in December 1985. The plume of steam is the result of lava pouring into the sea.
Left: View of the summit craters.

VULCANO, ITALY

Lat. 38.40 N – Long. 14.96 E – Alt. 1,640 feet (500 m) – 0101-05

4

Island arc

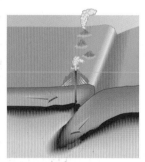

Stratovolcano

Prevalent volcanic activity. Vulcanian, phreatomagmatic, Plinian, lava flows, fumaroles. **General.** Vulcano, the southernmost of the Aeolian Islands, is composed of two volcanic centers that have been active in historical time: the cone of Fossa and Vulcanello, the youngest part of the island. The geology of the island can be divided into four main volcanic structures. Ancient Vulcano, which forms the southern end of the island, is a regular stratovolcano cut off by a level plain called the Vulcano Piano. This plain is believed to have been created following the partial filling in of a caldera structure, which swallowed the top of the ancient cone about 100,000 years ago. The cliffs that form the northwestern portion of the island (Lentia) are the result of layers of rhyolitic lava, emissions from about 15,000 years ago. The Fossa caldera is another structure resulting from a collapse. Located to the northwest of the caldera of the plain, it contains the cone of Fossa. Vulcanello, the northern peninsula of the island of Vulcano, is a lava platform on which stand three small pyroclastic cones. Vulcanello, originally a separate island, is believed to have risen above sea level in 183 B.C. Other eruptions, in the 6th and 16th centuries, led to the formation of the youngest pyroclastic cones and also a small lava flow. In addition, there is the large rock, really a modest cliff 214 feet (65 m) high, near the Porto di Levante, which is all that remains of an ancient pyroclastic cone whose products have been profoundly altered by fumarolic activity.

There are a great many descriptions, some dating back to Greek and Roman times, of eruptions in the crater of Fossa; the last eruption took place during the years 1888–90. The preceding eruptions had involved Plinian and phreatomagmatic explosive activity, with the production of numerous base surges and, to a lesser degree, lava flows. The most recent of such flows is that called Pietre Cotte, of which a tongue of rhyolitic obsidian remains, having descended along the northern slope up to the edge of the cone.

Until the end of the 1800s, sulfur and alum were extracted from the fumarolic area of the crater. The eruption of 1888–90 did not kill anyone but did devastate the surrounding territory, with an intense rain of bombs and blocks up to more than 3.3 feet (1 m) in diameter. This brought about the end of the mining activity.

Since 1980 the volcano has kept up its fumarolic activity, releasing steam and carbon dioxide, with lesser quantities of sulfur dioxide. The fumaroles usually have a temperature of around 212°F to 392°F (100°C to 200°C), but there have been instances of spikes in temperature above 1110°F (600°C)—in 1924 and 1993.

In 1985, 1988 and 1994 repeated earthquakes accompanied by swelling of the cone of Fossa along with changes in the quantity, temperature and chemical composition of the fumaroles awakened concern of a possible return to eruptive activity.

Opposite: The crater of Fossa with the fumarolic field.
Below: Aerial view of the cone of Fossa, with the town of Vulcano Porto.

Access and principal attractions. The climb to the crater of Fossa is an hour's walk along a well-marked path that sets off from Porto di Levante. Along the edge of the crater you can see fumarole fields and deposits of sulfur and large breadcrust bombs (so called for their shape after they hit the ground). Visible on the western side is the obsidian flow of the Pietre Cotte. By boat you can reach the rocky monoliths (the remains of ancient lava domes) between Vulcano and Lipari to the north. During the tourist season the Civil Defense office opens a visitors' center with information on many details of geology and the monitoring being carried out.

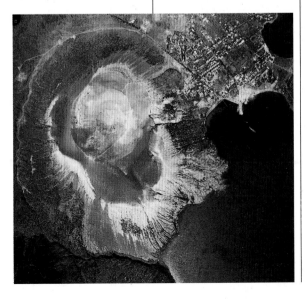

Shield volcano

Prevalent volcanic activity. Strombolian, lava flows, fumaroles.

General. With a base diameter of more than 22 miles (35 km) and an altitude of 10,990 feet (3,350 m), Etna is the largest active volcano in Europe. Its slopes are marked with many cinder cones, the result of lateral eruptions. On the summit are four active craters, called Bocca Nuova, Voragine, Northeast Crater and Southeast Crater, each with a diameter of 655 to 820 feet (200 to 250 m).

The most prominent structure is the Valle del Bove, a horseshoe-shaped depression about 2½ miles (4 km) wide, located on the eastern slope of the cone and open toward the sea. It was probably formed following an enormous avalanche that dumped a large part of the structure into the sea. The evolution of Etna took place in three main stages. During the first stage (300,000–150,000 years ago) there were large undersea eruptions of tholeiitic magma that created the deposits of pillow lavas today visible in the area of Aci Castello. During the second stage (150,000–80,000 years ago) a large shield volcano was formed of alkaline basalts and tholeiitic magmas. Finally, in the third stage various stratovolcanoes formed atop the shield volcano, made up of lava of varying composition (alkaline basalts, hawaiite, mugearite). Most of the eruptions that have taken place during historical time were from vents on the slopes of the volcano. The eruption of 1669 was particularly dramatic: on March 11 a fissure 6½ feet (2 m) wide and 30 feet (9 m) long opened in the southern flank of the volcano, causing weak eruptive episodes. On the same day a new fissure opened at a lower altitude, and lava was discharged a little to the north of Nicolosi, at only 2,625 to 2,790 feet (800 to 850 m) of altitude, followed by Strombolian-type activity that led to the formation of the cinder cones of Monti Rossi. Over the following days lava flowed to the south and reached Catania, flowing over the city's walls and destroying much of the city before

emptying into the sea.

When the activity finally ended, about 14¼ square miles (37 km²) had been buried beneath lava. Since 1669 numerous eruptions have occurred, reaching a frequency of more than one every two years over the past 25 years. The activity at Etna usually involves Strombolian phases followed by lava flows that extend from the upper parts of the volcano downhill, covering several miles. In some cases these flows have reached the towns along the slopes of the volcano, as happened in 1928, when the town of Mascali was destroyed.

In 1983, and even more so in 1992, the Civil Defense carried out two operations on

Access and principal attractions. From the Sapienza refuge (6,265 feet [1,910 m]) guided excursions depart for the summit craters, traveling first by cable car and then on foot (one hour climb); there is also an off-road van. The many attractions of Etna offer absolute beauty and endless scientific and naturalistic interest: the summit craters, the Valle del Bove (most of all the enchanting view of it from the summit of the volcano), the many eruptive vents, the hornitos (those formed in 1985 and visible beside the uppermost station of the cable car are very fine), the lava fields with their lunar appearance, the infinite number of lava flows in the most varied shapes and sizes and the lava tubes, including the Grotta dei Lamponi and the Grotto del Gelo, both of which can be visited at the foot of the volcano, the famous Gole dell'Alcantara with its torrent. The natural setting is wonderful, full of woods and local species of wildlife, protected as a park that can be visited on foot, by horse or on mountain bike.

Opposite: Spectacular view of the summit craters of the volcano.
Left: Computerized image of Etna, seen from the south. Visible is the immense Valle del Bove.

115

Below: Topographical map of the southeastern area of Etna, including the Valle del Bove, from the period of the 1989 eruption (from Barberi et al., Mt. Etna: The 1989 Eruption, CNR-GNV, Pisa, 1990). The area invaded by the lava flow is filled in red. The double broken line running from the southeast crater southward indicates the system of active fissures during the eruption. The green and blue arrows show the probable route the lava would have taken had it been discharged from the lower edges of the fissures, the most dangerous scenario. The little red triangles and lines mark the geodetic measurements made to take into account the physical deformations of the terrain.

Etna to arrest or deviate lava flows that threatened the town of Nicolosi and Zafferana Etnea. In December 1991, two fissures opened at the base of the Southeast Crater. One of these opened over 4,265 feet (1,300 m) along the southern flank of the Valle del Bove and, following a brief period of activity of lava fountains, a flow began at an altitude of 7,220 feet (2,200 m), pouring lava down the depression at an average rate of between 530 and 1060 cubic feet (15 and 30 m³) per second. It was clear that the lava would soon reach the small town of Zafferana Etnea, 5½ miles (9 km) farther down-

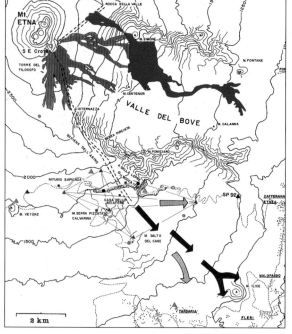

2 km

hill. Early in January 1992 an embankment was built near Portella Calanna, at an altitude of about 3,280 feet (1,000 m), and it held back the lava until April 10, when it overflowed the wall and in a few days reached the first houses of the town. At that point efforts were made at the upper area of the lava field with the aim of diminishing

the lava supply to the flow to make it slow or stop. These efforts culminated on May 27 when the stream of lava was deviated into an artificial canal at 6,560 feet (2,000 m) of altitude, making it explode its banks and pushing back the end of the lava flow many miles. The new flow covered only about 1.2 miles (2 km) before coming to a definite halt, marking the success of an operation that has remained unique in history: the manmade deviation of lava.

Above: Lava tube created during the 1991–93 eruption. Lava tubes form when the surface of a lava flow cools in contact with the air and solidifies to form a solid crust; such crusts can be many feet thick. Below the crust, lava continues to flow. When the eruption ends the lava drains downhill, leaving behind an empty tube, sometimes several miles in length.

Island arc

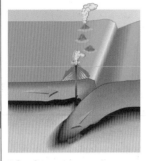

Caldera

Prevalent volcanic activity. Plinian, pyroclastic flows, lava flows.

General. Santorini, the ancient island of Thera, is one of the world's most beautiful volcanic islands. Its shape is the outcome of a large number of eruptions that have changed the geography many times over the course of millennia. The island is located at the center of the volcanic arc of the Aegean, produced by the subduction of the African plate under the Eurasian. The large island of Thera, the island of Therasia and the small island of Aspronisi are the remains of the outer walls of a large caldera, the rest of which is underwater, generated by a series of massive eruptions. The last of these, which took place in 1650 B.C., buried the city of Akrotiri, a civilization contemporary with the Minoans on the island of Crete and equally evolved. It also sank a large area to the north of today's island of Nea Kameni; the volume of material erupted has been estimated at about 7¼ cubic miles (30 km³). The eruption had an enormous impact on Hellenic civilization, and traces of it can be discerned in many works from antiquity; there are also many who see it as the basis for the myth of Atlantis, the ancient city famous for the glory of its civilization destroyed by a cataclysm and buried deep beneath the sea.

Following that large eruption, the volcanic activity on Santorini began at the center of the gulf and was of an essentially effusive character, leading to the gradual formation of a new eruptive center. The activity following the

Minoan-age eruption was at first underwater and eventually led to the emergence of the island of Palea Kameni in the years A.D. 46–47.

The activity then moved a few miles to the northeast and began to form the island of Nea Kameni, which began its emergence in the years 1570–73. Other eruptions, in the years 1707–11, 1866–70, 1925–28, 1939–41 and 1950, added new lava flows, significantly increasing the surface area of Nea Kameni.

All the eruptions of the last two millennia have been principally effusive, with only limited explosive activity of the Strombolian and phreatomagmatic types.

Access and principal attractions. An excellent road network makes movement easy on Santorini. The cliffs of this island, as well as those of Therasia, constitute a sort of large outdoor natural-history museum of the volcano. Facing in toward the caldera they display a dramatic succession of volcanic deposits that runs from flows of black andesite lava to layers of white dacite pumice, along with large sections of ignimbrite deposits from the 1650 B.C. eruption as well as those more recent. These cliffs form a sort of enormous cross-section view of the internal structure of the volcano, including many dikes. The island of Nea Kameni can be easily reached by boat. There are lava flows of andesite block lava as well as small craters produced by Strombolian activity. There is also the splendid view of Santorini, topped by the white buildings of the city.

119

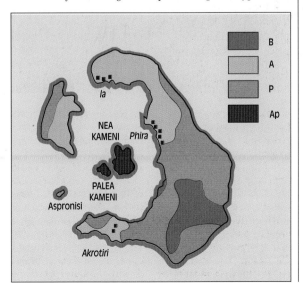

Opposite: Aerial view of the island of Thera from the south, with Capo Aspronisi in the foreground and the town of Phira on the plateau.
Left: Map of Santorini indicating the main geological formations: A, andesite; B, basalt; P, pyroclastic deposits; AP, andesite lava from recent eruptions.

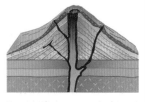

Island arc

Stratovolcano

Prevalent volcanic activity. Phreatic explosions.

General. The island of Nisyros lies between the islands of Kos and Rhodes at the eastern edge of the arc of active volcanoes in the Aegean Sea. The island is the above-water portion of an andesite volcano whose summit is cut off by a caldera of just under 2½ miles (4 km) in diameter, formed following a series of explosive eruptions that took place less than 24,000 years ago. Much of the caldera depression is occupied by a series of domes of dacitic and rhyolitic composition, the highest of which, Profitis Ilias, rises almost 1,970 feet (600 m) over the level of the caldera. Today numerous fumaroles at temperatures a little under 212°F (100°C) are present on the floor of the caldera. A geothermic probe carried out in the 1980s by the Greek national electric company revealed the presence of two aquifers beneath the caldera with temperatures of up to 572°F (300°C).

The southern portion of the caldera depression, known as the Lakki plain, and the small Lofos dome, were devastated in recent times by a series of phreatic explosions. In the period October–November 1871 a violent earthquake, followed by a series of explosions, occurred on the island. Red and yellow flames rose from the island, and fragments of rock flew out of the highest peak to fall on the island and into the sea. During the night two small phreatic craters were formed, Polybotes, in the central part of the depression of Polybotes Megalos, and another on the southern flank of the Lofos dome. The smoke discharged into the atmosphere wrapped the island in fog, while the bottom of the caldera was covered in ash.

Early in June 1873, following a series of violent earthquakes, a phreatic crater 20 to 23 feet (6 to 7 m) in diameter, called Phlegethon, opened on the southern side of the Lofos dome, near the edge of a fissure extending 165 feet (50 m) up to the Polybotes crater. For three hours the crater emitted a hot brine, together with blocks and fragments of rock along with fluid black mud.

On September 11 of that year an undersea fissure opened a few feet off the coastline near Mandraki; the seawater turned milky white, and a jet of vapor and hydrogen sulfide was observed for several seconds across the surface of the sea. On September 26 the craters Phlegethon and Polybotes gave off new emissions of brine, mud and blocks, becoming markedly wider.

Another phreatic event took place in 1887, far smaller than those preceding.

Access and principal attractions. *The island can be reached by boat from Piraeus, the port of Athens, three times a week year round. During the summer, it can be reached by boat from the island of Kos, with daily departures; from the islands of Symi and Rhodes there are three departures a week. You can tour the interior of the caldera by bus in excursions that last one to two hours. Or you can rent a scooter in the village of Mandraki and descend into the crater along a dirt path or a road that borders it. The caldera is an excellent example of a collapsed volcanic structure; the lava domes, fumarolic fields and phreatic craters inside it are of enormous interest to volcanologists. Also of great scientific value are the flows of rhyodacite perlitic lava on the southern side of the island and the caldera. The attractions of the island include the village of Mandraki, with its white and brightly colored houses, and the ruins of a Venetian castle at Emborios.*

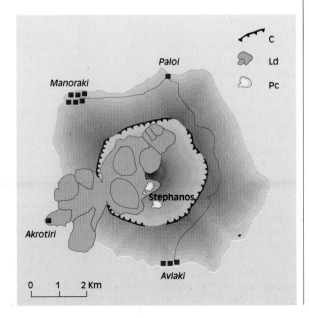

0 1 2 Km

Opposite: View of the interior of the caldera with phreatic craters and lava domes.
Left: Geological diagram of the island of Nisyros showing the borders of the crater (C), several of the phreatic craters (Pc) and lava domes (Ld).

PICO, AZORES, PORTUGAL

Lat. 38.47 N – Long. 28.40 W – Alt. 7,713 feet (2,351 m) – 1802-02 **8**

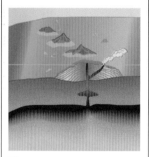

Hot spot

Stratovolcano

Prevalent volcanic activity. Surtseyan, Strombolian, lava flows, avalanches, Plinian.

General. Pico, the second largest of the nine Azores, is located at a point of conjunction of the North American plate to the west, the Eurasian to the northeast and the African to the southeast. The eastern area of the island is a plateau formed of basaltic lava at an average altitude of around 2,625 feet (800 m). Numerous cinder cones stand on this plateau, the highest of which (3,304 feet [1,007 m]) is Pico do Topo. The western side of the island, which was formed more recently, is a large stratovolcano, Pico. Its peak is 7,713 feet [2,351 m]

above sea level and another 3,280 feet (1,000 m) below, making it the highest point of the archipelago and indeed in all of Portugal. On the flanks of the volcano are more than 60 parasitic cones. The base of the volcano has very wide flanks with gentle slopes at angles between 5 degrees and 15 degrees. Around 3,940 feet (1,200 m) up there is a sudden increase in steepness where the voluminous central cone, made up of alternating layers of scoria and basaltic flows, begins; this is not only steep (30 degrees) but unstable, as indicated by the many accumulations in the area from avalanches. At the summit of the volcanic cone is a circular crater 655 feet (200 m) in diameter, with walls about 100 feet (30 m) deep, the bottom of which is occupied by an ancient lava lake that solidified about a thousand years ago. On the eastern edge of the crater is a small cone called Pico Pequeño (or Piquiño), 230 feet (70 m) high with very steep sides; at its base is a fumarole that discharges at 158°F (70°C).

The volcano has erupted five times since the 16th century. From 1562 to 1564 Pico do Cavaleiro, the cone located to the southeast side at 2,625 feet (800 m) in height, emitted lava flows that reached the ocean to the north near São Roque and Prainha. Early in February 1718, numerous eruptive vents opened along a fissure on the north slope of Pico at an altitude of 3,940 feet (1,200 m) above the towns of Santa Lucia and Bandeiras. The lava flows reached the coast, forming the present-day promontory of Cachorro; at the end of that month other vents opened in the south side at 985 feet (300 m) of altitude; for 10 months they gave off lava flows that reached and destroyed the village of São João.

In 1720 other vents opened at 1,312 feet (400 m) of altitude on the southeast flank of the volcano, near Soldão; the flows lasted five months. By the time the eruption ended a new cinder cone had been formed, known as Cabeco do Fogo.

Pico erupted in December 1963 on the northwestern tip of the coast.

Access and principal attractions. The ascent of Pico begins in the town of Furnas, following a path marked off with several signs. The climb does not present great difficulty unless one intends to scale the very steep flanks of Pequeño. Because of the variable weather conditions it is best to make the ascent in the company of a guide. The return trip toward the area of the port of Madalena offers the opportunity to explore the *"Furna dei Frei Matias"* lava tube, the interior of which has splendid basaltic stalactites. Many routes can be worked out to reach the eruptive vents of the volcano that have generated the enormous lava flows.

123

Opposite: The cone of Pico, seen from the island of San Jorge.
Left: Map of the volcano. The black spots indicate the lava flows of historical time.

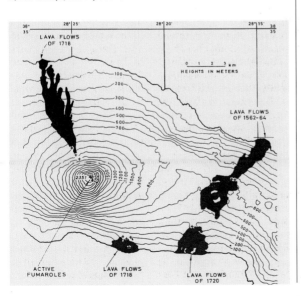

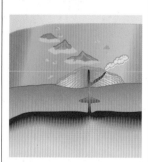

Hot spot

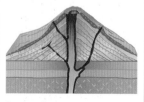

Stratovolcano

Prevalent volcanic activity. Plinian, pyroclastic flows, lava flows.

General. Agua de Pau is on the island of São Miguel, one of the Azores. The island lies across the Mid-Atlantic Ridge, on a topographical rise known as the Azores Platform, and consists of six volcanoes, three of which (Agua de Pau, Sete Cidades and Furnas) are active. Each of these volcanoes (as well as Povoacao, an extinct volcano on the eastern edge of the island) has a summit caldera and has experienced numerous explosive eruptions in its history that have covered the island with falling pumice and ash as well as ignimbrite deposits.

The oldest volcanic rocks of Agua de Pau are masses of trachytic lava (domes) that were erupted 180,000 years ago. Since then the volcano has erupted many other times, in most cases magma of trachytic composition. A large explosive eruption took place between 46,000 and 26,500 years ago, during which the current outer caldera was formed. A second large-scale eruption 15,000 years ago caused the sinking that created the current inner caldera.

Other large eruptions took place circa 6750 B.C. and 4550 B.C., contributing to the shape of the caldera. Then the volcano erupted at least eight more times, the last in the years 1563 and 1564. Most of the volcanic rocks on the island are pyroclastic deposits, but also common are domes and flows of trachytic-composition lava and lava flows, cinder cones and tuff rings of basaltic composition. The basaltic rocks are found on the flanks of the three active volcanoes, erupted principally from lateral vents, and also along an extended fissure area, most of it located in the stretch of ground between Agua de Pau and Sete Cidades.

Opposite: Lagoa de Fogo, located inside the summit crater of Agua de Pau.
Below: Map of the island of São Miguel indicating the main volcanic structures.

25°10'W

Atlantic Ocean

Sete Cidades Volcano

Agua de Pau Volcano

Nordeste Volcano

– 37°45'N

Ponta Delgada

Povoação Caldera

Furnas Volcano

0   10
KILOMETERS

Access and principal attractions. To make an excursion on foot in the area of the volcano you must leave the main coast road, and before reaching Agua d'Alto turn onto a steep path, 4 to 4¹/₂ miles (6 to 7 km) long, that leads from Serra de Agua de Pau to the summit caldera, known as Lagoa do Fogo. The caldera is occupied by a beautiful lake with clear water surrounded by white beaches accessible to hikers. Lava flows from the eruptions of 1563 and 1564, along with numerous lava tubes, are visible along the slopes of the volcano.

Other attractions of the island include the large Sete Cidades caldera, 7¹/₂ miles (12 km) in diameter, which contains two splendid lagoons, the Lagoa Verde and the Lagoa Azul; there is the hydrothermal valley of the Furnas caldera, with a lake of the same name studded with solfataras, many of them used as natural stoves (on which the famous cozido is cooked); there is the Caldeira de Pero Botelho with its boiling mud; the numerous "miradouro" of volcanic peaks, which offer splendid views of the island, and finally there are the characteristic towns of the island itself.

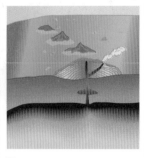

Hot spot

Shield volcano

Prevalent volcanic activity. Avalanches, fumaroles.

General. The island of El Hierro in the Canaries has a surface area of about 108 square miles (280 km²). The island, 165 million years old, is thought to be the above-sea section of a shield volcano that stands on the ocean floor at a depth of 12,140 to 13,125 feet (3,700 to 4,000 m).

The peculiar characteristic of the island is that it is located right on top of the hot spot of the Canaries. That aspect is made apparent in the characteristic triangular shape, a result of the presence of three rifts located at 120-degree angles where the volcanic activity is concen-trated. The three rifts form an equal number of morphological alignments running to the northwest, the northeast and the south.

Although only one eruption is known to have occurred during the last few centuries (in 1793, from the Lomo Negro volcano), El Hierro has the archipelago's greatest concentration of recognizable eruptive vents. These vents are located in accordance with the island's three morphological alignments.

The very wide depressions of El Julan to the southwest, Las Playas to the southeast and El Golfo to the north, continue onward several miles into the open sea. These depressions are composed of basins bordered by slopes in large part underwater and are believed to be the result of enormous avalanches from the volcanic structure. Much of the stratigraphy of the island can be seen along these slopes. Much has been learned about the internal structure of El Hierro by way of the many caves dug into it to exploit the island's subterranean water resources.

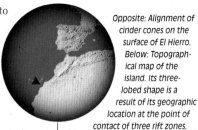

Opposite: Alignment of cinder cones on the surface of El Hierro. Below: Topographical map of the island. Its three-lobed shape is a result of its geographic location at the point of contact of three rift zones.

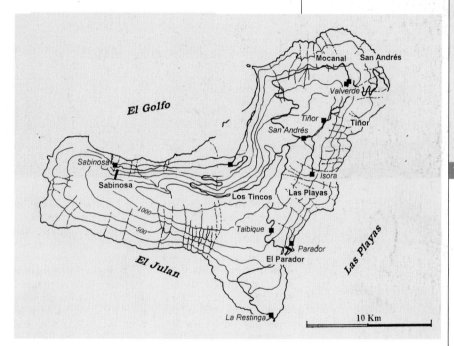

El Golfo

Mocanal San Andrés

Valverde

Tiñor Tiñor

San Andrés

Sabinosa

Sabinosa

Isora

Los Tincos Las Playas

1000

500

Taibique

Parador

El Parador

Las Playas

El Julan

La Restinga 10 Km

Access and principal attractions.

Valverde, the main city of El Hierro, is well located to serve as a starting point for excursions to every corner of the island. The main attractions include the many eruptive vents and lava flows, the great wall of El Golfo, built into the south side of an ancient volcano crater that collapsed into the sea, and Sabinosa, with its sulfuric thermal springs. Near Guarazoca you can visit El Hoy de Los Muertos, a necropolis dug into the rock containing embalmed bodies and funeral furnishings brought to light following an avalanche down the volcano's slope. Other attractions for visitors are the rock engravings of El Julan; the immense canary pine woods of Pinar; the belvederes of Jinama and Rincon, which offer splendid panoramas of the island; and the small, typical villages, in particular that of Guinea, where the homes are dug into lava tubes.

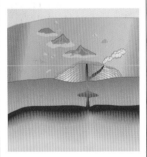

Hot spot

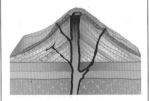

Stratovolcano

Prevalent volcanic activity. Lava flows, Strombolian, Plinian, fumarolic.

General. Tenerife, whose name in the ancient Guanche language means "snowy mountain," is the largest island of the volcanic archipelago of the Canaries: 52 miles (84 km) long and 31 miles (50 km) wide. It is triangular in shape with a surface area of 793 square miles (2,053 km²). The formation of the island began around five million to six million years ago as the result of a series of fissure eruptions of basaltic lava; later trachytic and phonolitic eruptions began forming a central volcanic cone about 9,845 feet (3,000 m) high. Following a series of explosive

eruptions that occurred sometime in the last 1.6 million years, the summit of the cone collapsed, forming the large caldera of La Cañadas, oval in shape, and about 10½ miles (17 km) long, with a perimeter of 47 to 50 miles (75 to 80 km) and walls 1,640 feet (500 m) high. During the last explosive eruption the northern slope of the island slid into the sea. About 200,000 years ago two stratovolcanoes began taking shape inside the caldera: Pico Viejo (9,797 feet [2,986 m]) and Pico de Teide, the highest point on the island (12,188 feet [3,715 m]), and also the highest peak in Spain.

Numerous historical eruptions of the volcano are known. Among the most recent and important is that of 1705, which took place on the northeast slope and began in a preexisting crater; during this eruption the Guimar was formed, a cinder cone 295 feet (90 m) high. In the same year another volcanic vent came into being, Fasnia. It gave off lava flows that spread up to 4 miles (6 km) away. The next year another vent opened, named Garachico, this time on the northwestern flank of the volcano. The river of lava it produced reached the sea, destroying the village of Tanque and the port of Garachico along the way. In 1798 the cone of Chahorra was built on the western flank; its lava flows stopped at the walls of the large caldera. The most recent eruption dates to 1909 and took place at Chinyero, a cone that formed on the western slope of the volcano and caused small explosions and lava flows.

Access and principal attractions. The ascent of Pico de Teide requires one day's time; 4 miles (6 km) outside Portillo you take a path for 3 miles (5 km) to a point where you rent mules for the climb to the Altavista refuge (10,695 feet [3,260 m]); from there another two hours' climb takes you to the peak. If you want to stay, reservations in the refuge require a permit issued by the island tourist office in Santa Cruz de Tenerife. The peak of Teide offers a splendid panorama of both the island and the caldera of Las Cañadas. There is also a cable car that will take you to 11,647 feet (3,550 m). After visiting the peak of Teide, do not overlook the interior of the caldera, which is worth seeing because of its lunar landscape (several science-fiction films have been made there), with an enormous variety of lava structures, including dikes that cross the walls.

129

Opposite top: The volcano of Tenerife atop which is the caldera of Las Cañadas.
Opposite bottom: Pico de Teide, 12,188 feet (3,715 m).
Left: Computer reconstruction of the island's topography. The summit caldera of Las Cañadas is clearly visible.

130

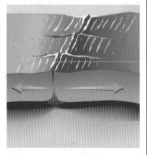

Oceanic ridge

Stratovolcano

Prevalent volcanic activity. Strombolian, hydromagmatic, lava flows.

General. The top of Snaefellsjökull is covered by a glacier nearly 4½ square miles (12 km²) in area; hence its name, which means "volcano with the snowy hood." It is located on the western tip of the Snaefells peninsula on a tongue of land that measures 18½ by 50 miles (30 by 80 km) extending from the western side of Iceland toward the Atlantic Ocean. The peninsula is host to numerous volcanoes, many of them extinct. Snaefellsjökull stands out among these because of its nearly 4,920 feet (1,500 m) of altitude, which make it vis-ible from a distance of more than 62 miles (100 km). From the geological and volcanological points of view, the region differs from the rest of Iceland; in fact, this is not a rift zone, and the composition of the magma erupted is not tholeiitic but rather alkaline, with products that, as in the case of Snaefellsjökull, sometimes reach high levels of chemical evolution. Jules Verne's *Voyage to the Center of the Earth* gave the volcano a certain fame since the main characters begin their trip by entering the Earth at the Snaefellsjökull crater and end it by reemerging from the crater of Stromboli in the Aeolian Islands.

The volcano is 700,000 years old. Beneath its thick covering of ice is a vast, partially collapsed crater, and various other craters, giving off lava flows, are highly visible on its sides.

Indications of at least 20 to 30 eruptive episodes have been identified. According to radionuclide studies only some of these events date to around 6,000 B.C., a period during which the activity at the central crater and at the side vents was characterized by moderate explosions with frequent production of lava flows. The most recent known eruption took place from the central crater around 200 B.C. Since then the volcano has been dormant.

Access and principal attractions. *Paths to the volcano leave from Olafsvik to the north and Arnarstapi to the south, about 130 miles (210 km) and 105 miles (170 km) from Reykjavik respectively. Two paths start at Arnarstapi and only one from Olafsvik, but it is the one most often used. From there the walk to the peak of the volcano takes about four hours, weather permitting; you must be prepared for all kinds of weather when attempting the ascent of a glacier. There is also a road leading up used by 4 wheel drive vehicles, and various organizations in town offer such excursions. You can also drive along Route 574, which goes around Snaefell-skökull; in the area of the volcano you will encounter numerous other volcanic structures to explore: to the south are the remains of the Londragar crater, located directly above reefs inhabited by large colonies of puffins; to the west is the cinder cone of Halohar Beludalur—the interior of which can be visited by car—and also the smaller cinder cone of Saxoll.*

EUROPE

Opposite and below: Views of Snaefellsjökull, a stratovolcano topped by a glacier. The mountain is part of a volcanic system running east-west over a distance of about 50 miles (80 km) on the Snaefells peninsula in northwestern Iceland.

131

SURTSEY, ICELAND

Lat. 63.18 N – Long. 20.36 W – Alt. 505 feet (154 m) – 1702-01

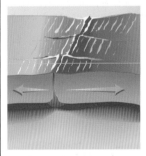

Oceanic ridge

Tuff cone

Prevalent volcanic activity. Surtesyan, Hawaiian, lava flows.

General. Between November 14 and 15, 1963, following countless and continuous eruptions, an island rose from the waters of the Atlantic Ocean in the Vestmannaeyjar (Westman Islands) to the south of Iceland. The Icelanders gave it the name Surtsey, from Surt, the giant fire god of Norse mythology.

The formation of the volcanic island involved three principal stages. The first stage was an undersea volcanic eruption through a fissure 1,805 feet (550 m) long that discharged large quantities of pillow lavas. These accumulated to form a volcanic structure that eventually rose more than 425 feet (130 m) to break the surface of the sea. During the second stage, which lasted roughly three months, the seawater had easy access to the eruptive vents, causing the rapid evaporation of the water in contact with the magma, and enormous columns of gray and white steam rose to heights of 6 miles (10 km), shaken by sudden explosions. During these explosions rock fragments (volcanic bombs) were shot into the air, leaving behind a black trail of volcanic ash that turned white as soon as the cooling steam condensed. By the end of

March 1964 this eruptive stage had led to the construction of an island about 1 mile (1.7 km) long, 492 feet (150 m) high, with a surface area of 0.19 square miles (0.5 km²), composed of loose volcanic detritus that would be easily eroded and washed away by the action of the sea. Such was not to happen, however, because other eruptions followed rapidly, increasing the surface area and cutting off direct contact between the water and magma. This also diminished the violence of the eruptions. This gave way to the third stage, beginning in April, which involved lava fountains and flows; the lava, far more resistant to erosion than the pyroclastic material

erupted in the earlier stages, formed a hard shell that guaranteed the survival of the island.

In June 1967, after three and a half years, the eruptions ended. They had generated an island 1 square mile (2.6 km²) in area. The observations made during the birth of Surtsey provided scientists with a wealth of new information on how volcanic islands are formed.

Access and principal attractions. Located at the far west of the Vestmannaey-jar, Surtsey is not an easy place for a tourist to reach and in fact is really open only to scientists—and even they need the proper authorization. Excursion boats leave from the nearby island of Heimaey, and there are planes that take off from the airport of Reykjavik and make a two-hour tour of the archipelago. Such flights are usually quite expensive. Less costly is flying over the island in a private plane; these leave from the small village of Bacchi on the southernmost tip of Iceland itself, right opposite the Vestmannaeyjar. The flight over the island is an unforgettable experience and offers views of the many volcanic structures in the region, including Heimaey, the next volcano in this book.

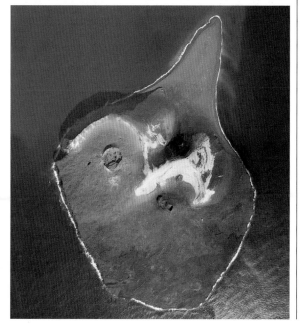

Opposite: Surtsey during the second stage of its formation, in the early months of 1964. The eruption that marked the birth of this island has given the name to a particular type of eruptive activity, called Surtseyan. It is a result of the interaction of low-viscosity magma and water. This produces violent explosions of steam with fragments of magma hurled into the air.
Left: Aerial view of the island of Surtsey today.

134

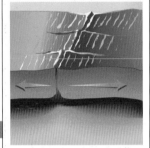

Oceanic ridge

Tuff cone

Prevalent volcanic activity. Hawaiian, lava flows.

General. Located 7½ miles (12 km) off the coast, the island of Heimaey is the most important maritime port for the Icelandic fishing industry. The volcanic complex that forms the island is the largest in size among those of the Vestmannaeyjar. Inactive for nearly 6,000 years, the volcano suddenly reawakened in the early hours of January 23, 1973. Along the eastern slope of the cone of Helgafell, located not far from the port, a fissure burst open and spread rapidly, reaching across the entire island. An apocalyptic vision rose behind the town, with nonstop lava fountains spouting into the air, lighting up the sky and throwing incandescent lava hundreds of feet into the air along the entire length of the fissure. The eruptive activity soon concentrated in an area of about 2,625 feet (800 m) to the northeast of the cone of Helgafell, where over the span of two days a 330-foot (100-m) high cinder cone was created, today called Eldfell. The rain of ash during the initial phases of the eruption caused serious damage to homes in the town; after that, the primary danger was from lava flows, which destroyed the northeastern area of the town and threatened to cut off the port. To prevent this from happening, large-scale efforts were undertaken to stop the lava or alter the speed at which it was advancing; most of all, enormous quantities of seawater were pumped onto it to cool it. After two weeks of incessant work the port was saved from closing. The eruption ended in the first week of June, after about five months of activity, by which time the new cinder cone stood 655 feet (200 m) high.

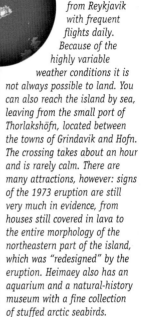

Access and principal attractions. The island of Heimaey can be reached from Reykjavik with frequent flights daily. Because of the highly variable weather conditions it is not always possible to land. You can also reach the island by sea, leaving from the small port of Thorlakshöfn, located between the towns of Grindavik and Hofn. The crossing takes about an hour and is rarely calm. There are many attractions, however: signs of the 1973 eruption are still very much in evidence, from houses still covered in lava to the entire morphology of the northeastern part of the island, which was "redesigned" by the eruption. Heimaey also has an aquarium and a natural-history museum with a fine collection of stuffed arctic seabirds.

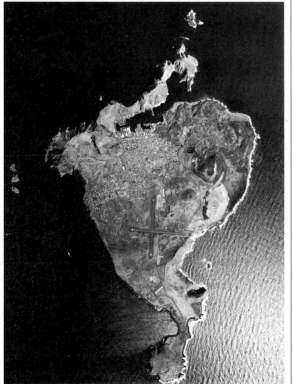

Opposite: A wall of fire rises behind Heimaey during the enormous 1973 eruption.
Left: Aerial view of the island. The darker parts in the northern area are the cinder cone of Eldfell and the lava flows from 1973. The eruption can be said to have had a beneficial effect on the island since before 1973 the northern area did not extend as far and the port was more exposed to the dangers of Atlantic storms. The lava flows made the inlet deeper, offering protection to the port and the island's large fishing fleet.

HEKLA, ICELAND

Lat. 63.98 N – Long. 19.70 W – Alt. 4,892 feet (1,491 m) – 1702-07

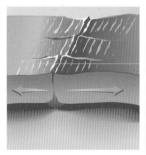

Oceanic ridge

Stratovolcano

Prevalent volcanic activity. Plinian, Hawaiian, lava flows, lahars.

General. Iceland's most famous volcano, known since antiquity as "the gateway to Hell," Hekla is located in the southern part of the island, 68 miles (110 km) east of Reykjavik. The volcano is composed of a ridge, 17 miles (27 km) long and 1½ miles (2.5 km) wide, crossed longitudinally by a fissure along which numerous craters are aligned. The eruptive behavior of Hekla is very particular: the volcano produces magma of both rhyolitic and basaltic composition and does so in eruptions that range from Plinian to lava fountains along very exten-

sive fissures. The volcano's physical structure reflects this double behavior, with elements that belong to stratovolcanoes mixed with those most often found in shield volcanoes.

The earliest historical eruption of Hekla dates to 1104. From then on, more than 15 major eruptions are known to have occurred, the intervals between them marked by many others of lesser size, generated from both the main fissure and from others in the surrounding area. About 12¾ square miles (33 km²) of magma are estimated to have been erupted during the past 10,000 years, making Hekla the most productive volcano in Iceland. The major eruption of the last century took place in 1947–48. On the morning of March 30, 1947, a yellowish brown cloud was seen to rise over the volcano and was soon followed by earthquakes. Immediately after this, a 3-mile (5-km) long fissure opened along the ridge of Hekla, giving off clouds of gas and ash along its entire length and melting snow, thus setting off violent flooding. The eruption cloud soon reached an altitude of about 12½ miles (20 km) in the atmosphere, obscuring the surrounding area and raining pumice and ash over almost all of Iceland; ash also fell on Finland. This initial phase of the eruption was soon followed by the abundant emission of lava flows, first along the entire fissure and then from the summit crater and a few other vents in the fissure, finally from the far end of the fissure at a height of about 2,720 feet (830 m). From that moment on, the volcanic activity came almost exclusively from Lava Crater, at the southwestern end of the fissure. It lasted about 13 months, with the total volume of material erupted estimated at about 0.29 cubic miles (1.2 km³).

The most recent eruptions of Hekla, far more minor in terms of volume of material erupted but still considerable in terms of violence, occurred in 1991 and 2000.

Access and principal attractions. *Hekla is located in the south of Iceland behind the Myrdalsjökull glacier. From Reykjavik you go south on Highway 1, known as the Ring Road, past the town of Selfoss. A few miles before Hella you turn left onto a road that leads into the interior of the island. After about 12½ miles (20 km) you reach the information center of Hekla, a sort of museum and learning center with videos and books on the volcano. From there paths lead to the top of the mountain. Along the way you find yourself crossing a wild, uncontaminated volcanic landscape dominated by lava and eruptive fissures, with layers of pumice and ash, the whole of it ringed by eruptive vents of varying appearance and size.*

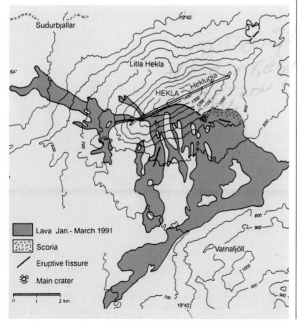

Lava Jan.- March 1991

Scoria

Eruptive fissure

Main crater

Opposite: Lava fountains and flows during an eruption of Hekla. Left: Topographical map of the volcanic massif indicating the eruptive fissures, the area affected by the lava flows of 1991 and the position of the main crater from which the flows originated.

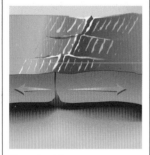

Oceanic ridge

Caldera

Prevalent volcanic activity. Strombolian, hydromagmatic, lahars.

General. Grimsvötn, in popular legend "the home of giants and trolls," is made up of a composite caldera completely covered by the Vatnajökull glacier, the largest in Europe, which with its 3,205 square miles (8,300 km²) of surface area hosts numerous other active volcanic centers. The caldera itself has a total surface area of about 13½ square miles (35 km²), and below it, beneath 495 to 820 feet (150 to 250 m) of ice, is a lake whose waters are kept warm by the constant flow of heat produced by the volcano's geothermal system. With a level of power equal to several thousand watts, this is fed by a magma chamber and a system of dikes that rise toward the surface from the chamber. Under normal conditions, steam is discharged from various points on the surface of the glacier. The melting of the glacier and the arrival of fluids from the geothermal system of the volcano supply water to the lake at the rate of about 706 cubic feet (20 m³) per second. The level of the lake thus progressively increases, raising the covering of glacial ice over it and producing an increase in pressure. When this pressure reaches a critical level, the base of the glacier begins melting, forming a subglacial river. This river can flow more than 31 miles (50 km) under the glacier, right up to its far end, where it empties into the Skeidara River in the Skaftafell national park, about 12½ miles (20 km) from the coast and less than 3 miles (5 km) from the Ring Road, Iceland's main artery. The amount of water flowing along this subglacial river generated by the oscillations of the surface of the caldera lake varies from several thousands to tens of thousands of cubic meters per second, causing gigantic floods that in many cases have swept away or destroyed everything in their path.

In addition to these floods there are those caused by the frequent eruptions of the volcano, considered the most active in all of Iceland, with 40 to 50 eruptions known or at least believed possible in historic time. Such eruptive

events cause the rapid melting of large masses of the glacier, resulting in outburst floods known as *jökulhlaup* that can be true natural catastrophes.

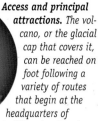

In October 1996 an eruption discharged about 0.17 cubic miles (0.7 km³) of lava. An estimated 0.8 cubic miles (3.4 km³) of meltwater resulted, and on November 5 it caused a flood of water mixed with mud and ash that ranks as the worst of the past 60 years in terms of the amount of material involved and the speed at which it moved, destroying many miles of road as well as bridges in the region to the south of the large glacier. There was further eruptive activity in December 1998.

Access and principal attractions. The volcano, or the glacial cap that covers it, can be reached on foot following a variety of routes that begin at the headquarters of Skaftafell national park, which is about 248 miles (400 km) from Reykjavik. Hofn, a town to the northeast of the park, is the base for excursions across the ice itself in 4 wheel drive vehicles, snowcats and powerful motorized sleds. There are many reasons to visit the immense glacier of Vatnajökull, both in terms of the area's landscape and its nature. For example, a 30-minute walk from the park headquarters reaches the enchanting Svartifoss ("black waterfall"), famous for its waterfall amid spectacular cliffs formed of columns of basaltic lava.

Opposite: View from the southern coast of Iceland of the Vatnajökull glacier, under which are numerous volcanic structures.
Left: The fissure along the surface of the glacier during the subglacial eruption of 1996. It is 4 miles (6 km) long and 655 to 1,640 feet (200 to 500 m) wide.

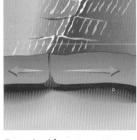

Oceanic ridge

Fissure

Prevalent volcanic activity. Strombolian, Hawaiian, lava flows.

General. The eruptive fissure of Lakagigar, also known as Laki, from the name of an ancient volcanic structure located along its axis, is made up of an alignment of about 130 eruptive craters spread out over a distance of about 15½ miles (25 km) running northeast-southwest, thus following the main tectonic alignment of Iceland. These craters were formed during a volcanic event of colossal dimensions in the years 1783–85, considered the worst natural catastrophe in the entire history of Iceland. Volcanic activity began along the fissure with small phreatomagmatic explosions that were followed by Strombolian explosions and finally repeated effusive phases. About 3 cubic miles (12 km³) of magma were erupted, of which 0.18 cubic miles (0.75 km³) were made up of tephra and the rest basaltic lava of tholeiitic composition. The lava was discharged from craters at a rate of 10,465 cubic yards (8,000 m³) per second and spread over almost 232 square miles (600 km²) of territory. Along with the solid materials, 500 million tons of gas were discharged into the

atmosphere. The eruption, which ranks as the worst lava flood in history, had catastrophic consequences for the Icelanders. In the following months a bluish gas rose from the solidifying lava and was carried across the entire island by winds, poisoning much of the pastureland and cultivated crops. More than half the livestock were killed by poisoning or by hunger, beginning a long period of starvation that led to the deaths of nearly 9,000 people, at the time equal to 20 percent of Iceland's population.

Access and principal attractions. *The spectacular alignment of Laki craters is located to the immediate southwest of Vatnaökull glacier. The best way to reach the area is to start off at Kirkjubaejarklaustur, about halfway along the Ring Road between Reykjavik and Hofn, and travel to Kirkju, a small town with a population of 300 that has a hotel and a campground. Just outside the town is a dirt road, demanding in any weather and open only in the summer, that leads to the cone of Laki. You will need a well-equipped off-road vehicle to get over the many fords and make it across the lava fields along the route without damage. A path leads to the peak of the volcano, which can be reached in less than an hour. From there you can see the spectacular alignment of small craters marking the Lakagigar fissure.*

Opposite: The spectacular alignment of cinder cones along the Lakagigar fissure in the tectonic trench of Skaftar.
Left: Aerial photograph showing the entire length of the Lakagigar eruptive fissure, a distance of about 15 miles (24 km). To the northeast the fissure continues beneath the Vatnajökull glacier for another 28 miles (45 km) to intersect the caldera of Grimsvötn.

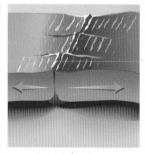

Oceanic ridge

Caldera

Prevalent volcanic activity. Strombolian, Hawaiian, lava flows, Plinian.

General. Located in the central part of northern Iceland, in one of the country's most remote regions, the volcano complex of Askja is composed of three interconnected calderas with a total surface area of 17⅓ square miles (45 km²). They are located at the center of a volcanic massif, composed primarily of hyaloclastites, the Dyngjufjoll, which with a height of around 4,920 feet (1,500 m) rises off an immense expanse of tholeitic lava fields known as the Lava of Evil Deeds.

The volcanic complex of Askja has erupted about 40 times over the last 10,000 years, both from the area currently occupied by calderas and from the numerous surrounding fissures.

The first historical eruption took place in 1875. In February 1874 a violent earthquake and the discharge of a large column of steam from the volcano announced the imminent return of eruptive activity. These phenomena continued throughout the year. In January 1875 basalt lava began flowing from fissures to the northeast and south of the volcano. On March 28 and 29 of that year, explosive eruptions expelled about 0.12 cubic miles (0.5 km³) of pumice of rhyolitic composition and a lesser amount of rhyolitic pumice mixed with basalt. The eruption caused the further collapse of part of the caldera's base over an area of roughly 4¼ square miles (11 km²). The new depression is today filled by a 985-foot (300-m) deep lake. The most recent eruptions, all of them smaller in size and involving for the most part lava flows and fountains, happened in the 1920s, in 1938 and in 1961, when a new ring-shaped eruptive fissure opened on the northeast edge of the caldera, producing spectacular lava fountains.

Opposite: The lake that occupies the caldera was formed following the 1875 eruption. Its depth of 985 feet (300 m) makes it Iceland's deepest lake. Below: Diagram showing the activity beneath the volcano. A series of dikes rises from the magma chamber to feed both the caldera and the many eruptive fissures in the area.

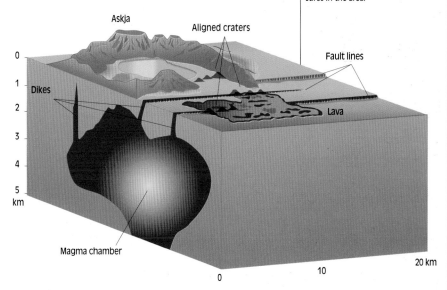

Askja

Aligned craters

Fault lines

Dikes

Lava

Magma chamber

0
1
2
3
4
5
km

0   10   20 km

Access and principal attractions. The complex of calderas at Askja is in a wilderness area region accessible only to off-road vehicles, and even then only under good conditions. The would-be tourist must bear in mind that there are many fords and accumulations of sand and ash to be passed before reaching the site. Taking the Ring Road from the town of Seydisfjordur,

you reach Eglisstadir, from where you proceed north on the same road until you approach the small town of Modruladur. Just before the town is a fork in the road with signs indicating Askja and the roadway that crosses the island. From that point it's still 62 miles (100 km) to the footpath. An hour's walk brings you to the edge of the caldera. The landscape of Askja offers many

interesting structures, from the large composite caldera, partially occupied by the lake, to the main eruptive fissures and lateral cones. This is one of the most desolate and fascinating areas of Iceland. During the 1970s, the area was used by NASA astronauts for special exercises and tests.

144

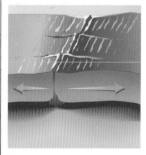

Oceanic ridge

Caldera

Prevalent volcanic activity. Hawaiian, Strombolian, phreatomagmatic, lava flows, hydrothermal.

General. The Krafla volcanic system is part of a row of eruptive fissures that extends in a north-south direction across the northern area of the island. It constitutes one of the most active branches of the Iceland Ridge. The system of active fissures has an overall length of 56 miles (90 km) and a width of 2 to 5 miles (3 to 8 km). In the narrowest sense, Krafla volcano is a caldera measuring 5 by 6 miles (8 by 10 km) located a few miles to the north of Lake Myvatn. The interior of the caldera includes several cones and tuff rings, including Mount Krafla, a palagonite peak that gives its name to the caldera.

About 2,500 years ago, the Krafla volcanic system was the site of a major eruption. During this episode a series of fissures at least 5 miles (8 km) in length came into being. The two main fissures, called Lúdensborgir and Prengslaborgir, both marked by the perfect alignment of various cinder cones, erupted a large quantity of lava that accumulated in a depression and blocked off the flow of nearby water courses, causing the creation of Lake Myvatn.

In 1724 a period of eruptions began that lasted until

1729 and led to the formation of seven fissure vents. Initially these were located inside the caldera itself, but they progressively extended outside its borders, covering a surface area of 12¾ square miles (33 km²) with new lava. During the initial phase, violent phreatomagmatic explosions took place inside the caldera, leading to the creation of the maar of Viti. In the years between 1975 and 1984 a series of eruptions took place from the fissure system to the north of the Krafla caldera. During these eruptions, much liquid lava was emitted in flows and fountains hundreds of feet high. By the time this eruptive period ended the group of fissures was 33 feet (10 m) wider. Today, the Krafla volcanic area is full of geothermal activity. The hot fluids released by the magma chamber located at a depth of 2 to 4½ miles (3 to 7 km) are extracted from the underground, supplying the Krafla Geothermal Power Station, located near Lake Myvatn, with 30 megawatts of power. The station supplies energy and heat to homes and greenhouses.

Access and principal attractions.
From Reykjavik, the Krafla-Myvatn area can be reached by driving west on the Ring Road. Another way, more difficult and adventurous, is to leave from the south and cut across the large glaciers and the uninhabited zone of Sprengisandur. You'll need to be well equipped with a suitable off-road vehicle and a good supply of gas. The area of Lake Myvatn is considered a true natural paradise and offers one of the rare examples of explosive volcanism in Iceland, with a large number of craters formed by the interaction between lava and the lake water. Lake Viti is a beautiful example of a maar. The eruptive fissures of Lúdentsborgir and Pregslaborgir bear rows of numerous cinder cones.

Opposite and left: Two views of the maar of Viti in the Krafla volcanic area. Maars are shallow, flat-bottomed, usually circular craters; most fill with water, forming natural lakes.

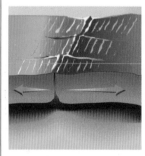

Oceanic ridge

Table mountain

Prevalent volcanic activity. Lava flows, phreatomagmatic activity.

General. Since this mountain is not known to have erupted even once during the last 10,000 years, it cannot be considered an active volcano. It is included in this book because it is a perfect example of a table mountain and because the processes that led to its formation are believed to be the same as those of the active volcanoes today located under the Vatnajökull volcano.

Structures like this are usually the result of eruptive fissures of only limited length. The formation of such structures is divided into the four principal phases illustrated in the diagrams on the opposite page.

1) Magma begins to rise toward the surface through a fissure. The increase in temperature generated by this volcanic activity melts part of the glacial ice above, creating a large subglacial sack of water and steam. The encounter between the rising magma and the meltwater from the glacier results in the formation of pillow lavas.

2) The uppermost part of the accumulated pillow lavas, in contact with cold water, is transformed into an aggregate of glass shards called hyaloclastites. These undergo a further alteration into a

yellowish claylike material called palagonite. At this point both pillow lavas and palagonite are present at the base of the glacier. If the accumulation of volcanic material is sufficient, the pressure of the water above it diminishes until it reaches a critical point at which hydromagmatic explosions begin.

3) The accumulation of pillow lavas and palagonite directs the further discharges of magma toward the surface instead of under the layer of water; these lava flows cover the previously erupted material.

4) The glacier is eventually melted by climactic changes, leaving exposed the underlying table mountain with its flat summit and very steep sides made up of palagonite and pillow lavas.

Access and principal attractions. To reach the table mountain of Herbubreid from Reykjavik you must take the Ring Road all the way to the other end of the island and the area of Lake Myvatn. From there you need an off-road vehicle to take the road F-88 heading south toward the area of the Askja volcano. After 37 miles (60 km) you will pass a small airport and campgrounds; Herdubreid will be visible to the right. Another route, more difficult and adventurous, involves heading south through the large glaciers and crossing the desert area of Sprengisandur, but you will need to be well equipped with an off-road vehicle, plenty of gas and heavy clothes for the cold. The region of Herbubreid is interesting for many reasons. Aside from being home to Iceland's prototypical table mountain, there are the formations of pillow lavas and palagonite and, within a radius of 62 to 93 miles (100 to 150 km), there is the area of Lake Myvatn and the Askja and Krafla volcanoes.

Opposite: The table mountain of Herdubreid stands on a lava plain in a desolate corner of Iceland. Below: Diagrams of the main phases typical of the formation of table mountains. Magma rises beneath a glacier, melting part of it and interacting with the meltwater, building pillow lavas, until the eruptive vent makes contact with the atmosphere. This leads to subaerial lava flows. The later melting of the ice cover leads to the exposure of the table mountain.

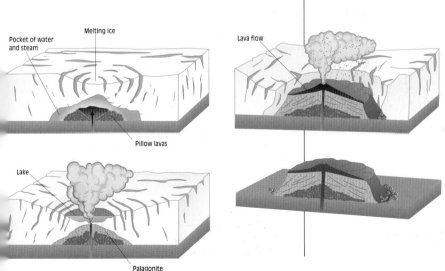

Melting ice

Pocket of water and steam

Pillow lavas

Lava flow

Lake

Palagonite

The African continent is made up in large part of rocks that are billions of years old in a structural arrangement that approximates a state of balance between the cold, solid rocks of the surface the warm, fluid rocks of the underlying mantle. This structural arrangement is apparent in the continent's average height above sea level of about 3,280 feet (1,000 m) and in its geography, which is dominated by vast plateaus. Even so, geological phenomena are taking place inside the African continent, even in our own time, that are changing the appearance of the land and doing so at such great speeds that large changes take place over spans of tens of millions of years, a relatively brief time in the geological history of our planet. These phenomena are connected to the opening of an oceanic ridge in the eastern corner of Africa, and they manifest themselves as a series of fissures along with extensive modification of the terrestrial morphology. This area is known as the Great Rift Valley, or the East African Rift Zone.

Volcanoes of the East African Rift Zone

The system of the African Rift begins at the point of convergence of the Red Sea (which has two above-sea-level volcanoes) and the Gulf of Aden (one above-sea-level volcano) in the area around Djibouti (four volcanoes); from there it crosses Eritrea and Ethiopia (47 volcanoes) toward Kenya (21 volcanoes), where it splits into an eastern branch that crosses Tanzania (10 volcanoes) to reach Mozambique and a western branch that crosses Uganda, Congo, Rwanda and Burundi (14 volcanoes in all). The total length of this Great African Rift is about 6,835 miles (11,000 km), almost one-fourth of the circumference of the Earth. The region of the rift forms a narrow strip with more or less clearly discernible branches; the rocks of the mantle beneath that region are less dense that those in the surrounding areas and therefore tend to rise due to buoyancy forces. This rise produces pressure on the overlying crust, which arches and narrows, producing the morphology typical of a rift, a long depressed zone bordered by deep slopes. Water flowing down the walls of the rift gathers in these depressions, forming large lakes, such as Turkana in Kenya and, farther south Lakes Tanganyika and Nyasa (Malawi), while Lake Victoria is located within the cratonic area of Tanganyika, bordered by the eastern and western branches of the rift. Seawater can make its way into such depressions, as in the case of the Red Sea and the Gulf of Aden, leading to the formation of true small oceans with their own midocean ridges.

The movements of the crust that are causing the East African Rift began about 25 million years ago, and since then they have been moving ahead at an average rate of expansion of about 0.04 inches (1 mm) per year. Numerous volcanoes have come into being as a result of the movements involved in the rise of the mantle and the narrowing of the crust, and a chain of volcanoes is located along the entire length of the rift. Over the

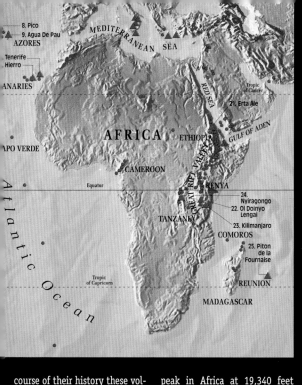

Labels on map:
8. Pico
9. Agua De Pau
AZORES
MEDITERRANEAN SEA
Tenerife
Hierro
CANARIES
RED SEA
Tropic of Cancer
21. Erta Ale
AFRICA
ETHIOPIA
GULF OF ADEN
CAPO VERDE
CAMEROON
Equator
KENYA
24. Nyiragongo
22. Ol Doinyo Lengai
TANZANIA
23. Kilimanjaro
COMOROS
Atlantic Ocean
25. Piton de la Fournaise
Tropic of Capricorn
REUNION
MADAGASCAR

Volcanoes in the western Indian Ocean

In the Mozambique Channel between the African continent and Madagascar (five volcanoes) are the Comoro Islands (two volcanoes), where Karthala showed explosive activity in 1991. About 435 miles (700 km) to the east of Madagascar is the island of Réunion in the Indian Ocean, site of an extremely active volcano Piton de la Fournaise, which over the past 50 years has erupted an average of every 14 months, with prevalently effusive activity and deposits of basaltic lava flows.

The volcanic chain of Cameroon and the volcanoes of Libya, Chad and Sudan

Another area of volcanic activity is in Cameroon, on the western side of Africa, site of an intercontinental hot spot that has generated several volcanoes, the most active of which in historic times is Mount Cameroon, which last erupted in 1982. Other active volcanoes located within the African plate are in Libya (one volcano), Chad (four volcanoes) and Sudan (five volcanoes). None of these is known for certain to have erupted during historical time.

course of their history these volcanoes have seen both explosive and effusive activity, the composition of the magma being for the most part alkaline. Aside from the area of active volcanoes located on the bottom of the Red Sea and the Gulf of Aden, the volcanic system of Erta Ale in the Afar region of Ethiopia, near the area of confluence of the three main branches of the rift, is of great importance. This volcano's traits mix those common to a hot spot with those common to a midocean ridge. Other important volcanoes are located along the branch of the rift that heads south from the Afar. These include Kilimanjaro, the highest

peak in Africa at 19,340 feet (5,895 m) above sea level; Ol Doinyo Lengai, in the Gregory Rift of Tanzania, the only known active carbonatite volcano in the world; and along the chain of the Virunga in the Congo, Nyamuragira and Nyiragongo, both of them active quite recently, the former with eruptive episodes in February 2001, the latter the scene of one of the few active lava lakes in the world and of violent activity as recently as January 2002. Most of these volcanoes are located in remote regions with few inhabitants and thus do not present great dangers to humans, but such is not the case with Nyiragongo

Continental rift

Shield volcano

Prevalent volcanic activity. Hawaiian, lava lake, lava flow, caldera collapse, fumarole.

General. The volcanic chain of Erta Ale, located in the north of Ethiopia, is part of the volcanic system associated with the formation of the Great Rift Valley, a depression about 6,835 miles (11,000 km) long produced by the processes connected with the formation of a new ocean. A few million years from now this new ocean will separate the eastern region of Africa from the rest of the continent, much as the Americas were separated from the European and African continents by the creation of the Atlantic Ocean about 60 million years ago.

The processes creating the rift are at a particularly advanced stage in the region of Erta Ale, such that the magma emitted by the volcano has the tholeiitic composition usually associated with magma discharged from an oceanic ridge. The fact that the ridge is emergent and thus directly observable by volcanologists makes this volcanic system of exceptional importance within the entire panorama of volcanology on our planet. Even so, because of its remote location, little reliable information was available concerning the volcano before 1967, when annual observations began. During the 1970s six Italian-French scientific expeditions went into the region, which until then had been largely unexplored, and greatly increased the available scientific knowledge. The enormous structure of Erta Ale is composed of an elongated shield volcano stretching more than 62 miles (100 km) in a direction parallel to the Red Sea; it is about 25 miles (40 km) wide. At least six active volcanic centers are located within that area. The center of Erta Ale is the fulcrum of the eruptive activity, with the perennial presence of one or more lava lakes within its large summit caldera. The caldera, oval in shape and 1.1 miles (1.8 km) long at its main axis, was created following a series of collapses. Inside it are two circular pit craters.

The largest of these (985 feet [300 m] in diameter) is the scene of intense fumarolic activity along its walls and floor and sometimes also a lava lake; the smaller pit crater (510 feet [155 m] in diameter with vertical walls about 265 feet [80 m] high) has a lava lake that has been active at least since 1967. The lava lake is like a window through which one can observe the process of magma rising directly from the mantle.

Opposite: The lava lake at Erta Ale. Below: The summit caldera. The two pit craters are visible in the upper part. The smaller pit crater has a lava lake.

Getting to Erta Ale is no simple matter. First of all the difficulty of its remote location on the Piana del Sale in Danakil, a border region between Ethiopia and Eritrea; then there are the difficulties presented by the physical realities of the place, with temperatures that exceed 122°F (50°C) and little available water; and finally there is the political situation. The territory is controlled by groups of Afar warriors with whom friendly relationships are difficult to establish. It is also the scene of actual battles: the relationship between Ethiopia and Eritrea is unstable, ranging from open war to tense hostility. Thus to reach Erta Ale requires the organization of a true expedition, without leaving anything to chance: all necessary permits and authorizations must be obtained well in advance, and up-to-date information on the political situation is essential. The most interesting aspect of the Erta Ale chain is the fact that it is an oceanic ridge; there is also the opportunity of observing active lava lakes, a phenomena visible, at varying times, in only three other volcanoes on Earth: Kilauea in Hawaii, Erebus in Antarctica and Nyiragongo in Congo. Large gas bubbles are constantly being discharged from the surface of the lake, and these produce lava fountains up to a height of several feet. At night the lake illuminates the sky like an enormous fiery torch pointed upward.

152

Continental rift

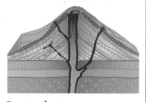

Stratovolcano

Prevalent volcanic activity. Lava flows, fumaroles, Hawaiian, Vulcanian.

General. The spectacular cone of Ol Doinyo Lengai, located 9½ miles (15 km) to the south of Lake Natron, rises almost 9,840 feet (3,000 m) from the broad plain of the Gregory Rift Valley. This highly unusual volcano is the only one in the world to have erupted carbonatite lavas, instead of silica, in historical time. In such lavas silica, the basic constituent of every other magma emitted on Earth, is practically absent.

These magmas are believed to have resulted from a process of chemical breakdown involving two liquids in a magma chamber, one of carbonatite composition, the other silicate. The breakdown would have resulted from the particular abundance of carbon and calcium in the initial liquid.

The summit of the volcano is composed of two craters, only one of which, the northernmost, is active. In recent times this activity has consisted of the effusion of carbonatite lava flows inside the summit crater accompanied by less frequent phases of small explosions similar to the Strombolian eruptions of silicate volcanoes. The interior of the crater is studded with a high number of hornitos and cinder cones in an extremely large variety of shapes and sizes.

The lava is emitted at temperatures of only 932°F to 1112°F (500°C to 600°C), about half of that of tholeiitic or alkaline lavas. Despite this, the highly unusual composition of the lava makes it many times more fluid than theolitic or alkaline basalt lavas, so the lava flows are sometimes only an inch or two wide, thus producing structures whose morphology makes them resemble miniature lava flows from silicate volcanoes. Aside from being extremely fluid, the lava emitted is as black as oil. However, as soon as it begins to cool, the process of chemical interaction with the atmosphere dissolves the carbonate, transforming its composition, at which point the lava becomes as white as snow. Therefore the crater's interior looks like a big white landscape crisscrossed in all directions by the black tongues of just-erupted lava. The historical activity of Ol Doinyo Lengai involves the alternation of long periods of effusive activity, which can last months or years, with shorter periods

of explosive activity of the Vulcanian type, which tend to last weeks or months. During the past hundred years, larger explosive eruptions took place in 1917, when lava descended along the slopes of the volcano, in 1940–41 and in 1966–67. During each of these eruptions volcanic ash and cinders were carried over distances of more than 62 miles (100 km) and deposited over vast areas, destroying vegetation because of their highly alkaline composition.

The volcano's cone is today almost bare of vegetation, with only patches of grass and a few bushes, but the first explorers to reach the site during the last century described it as covered with a dense, tropical vegetation that was in fact an obstacle to climbing the mountain.

Access and principal attractions. *From Arusha, the jumping-off point for safaris heading into the famous national parks of Manyara, Serengeti and Ngorongoso, you reach Mto Wa Mbu. From there you abandon the main road to head north toward Lake Natron, running along about 25 miles (40 km) of the western wall of the Gregory Rift, an enormous slope, 13,125 to 19,685 feet (4,000 to 6,000 m) deep. At the end you can see the cone of Kerimasi and behind it Ol Doinyo Lengai. Six miles (10 km) to the north is Lake Natron, nearby which are two areas for camping. The ascent of the volcano is somewhat challenging, given its steepness; at least four to six hours are needed, making a departure during the night necessary to avoid the hottest hours of the day. The interior of the crater offers a singular spectacle: many cinder cones and hornitos stand on a snow-white plain cut by many tongues of fluid black lava and by radial fissures out of which rise volcanic gases. The view from the summit of the volcano covers the great wall of the Rift Valley and the vast plains of the African savannahs, with many volcanoes in the area up to Lake Natron and Kenya.*

Opposite: The cone of the volcano rises abruptly off the savannah.
Left: In July 1996 the active crater of Ol Doinyo Lengai was almost completely full of carbonatite lava. In April 1999 lava flows discharged from the crater poured down the slopes of the volcano.

Continental rift

Stratovolcano

Prevalent volcanic activity. Lava flows, Strombolian.

General. Kilimanjaro is a particularly fascinating mountain. Its rounded silhouette, with the top covered by perennial snows, is the background to classic images of Africa, the foreground filled with herds of zebra or elephants.

The park of Kilimanjaro has an area of 292 square miles (756 km²) and includes the entire volcanic massif with its principal peaks, Kibo (19,340 feet [5,895 m], the highest in Africa), Mawenzi (16,893 feet [5,149 m]) and Shira (12,999 feet [3,962 m]). Kilimanjaro is one of the world's highest isolated mountains; it rises 15,750 feet (4,800 m) off the flat surrounding savannah. Kibo has a summit caldera 1.2 miles (2 km) in diameter; in it is a second caldera 2,690 feet (820 m) in diameter; inside that is a 395-foot (120-m) high cinder cone.

From a geological point of view, Kilimanjaro is far younger than generally thought; according to the most recent studies, its current shape dates back only 750,000 years.

The caldera of Shira is the oldest volcanic center, having ceased its activity about 500,000 years ago. Kibo and Mawenzi were once higher and have gone through periods of common activity. Mawenzi, originally taller than Kibo, ceased its volcanic activity after Shira did. The rocks that make up its peak have been broken up by the action of erosion, much like what happened to Shira. Only Kibo has preserved the morphology of its summit volcano.

The history of the exploration of Kilimanjaro is full of exciting episodes. Ptolemy wrote of a "great snowy mountain," and references to it show up in the chronicles of several Chinese geographers dating to the period of the Middle Ages.

In 1519 the Spaniard Fernandes de Encisco described the "Olympus of Ethiopia to the west of Mombasa." The first modern European to see it was the Swiss missionary Johan Rebmann, but his report of seeing, on May 11, 1848, an immense snowy mountain in the heart of the savannah at the level of the Equator was not believed. It was the German nobleman Karl Klaus von der Decken who put an end to the debate, and in 1861 he carried out the first attempt to climb to the peak. He failed to make it all the way up but brought back to Europe the first accurate description of the volcano. In 1887 another German, Hans Meyer, professor of colonial geography at Leipzig, cartographer and explorer, began a series of attempts that finally, in 1889, got him to the peak together with a mountain climber from Salzburg named Purtscheller. In 1938 a pair of mountain climbers from Stuttgart opened the first trail up the southern slope of Kibo. Despite the outcomes of the two world wars, the highest peak in Africa remained dedicated to the German kaiser until 1961.

It was D.V. Lathan who discovered the famous frozen leopard on the summit of the volcano, and Ernest Hemingway who made the discovery famous in his short story, "The Snows of Kilimanjaro."

Today one does not need to be a famous explorer or a skilled mountain climber to ascend Kilimanjaro. Adequate physical training and the right equipment are all that is needed. Thousands of fans make the climb every year.

Access and principal attractions. *Several routes lead to the top of Kilimanjaro. They vary in terms of the time required, the path and its difficulty. The Marangu path, which begins at the entry to the park (6,100 feet [1,860 m]), is the easiest and also the best equipped in terms of huts. Five days are needed for the excursion to the peak, of which one and a half are the descent. The hike through the tropical forest at the base of the mountain, the climb across the glacier and the endless panorama from the top attract thousands of tourists from around the world every year.*

Opposite: The snowy crater of Kibo, the highest peak in Africa (19,340 feet [5,895 m]).
Below: Map of the volcanic massif indicating the main routes to the top.

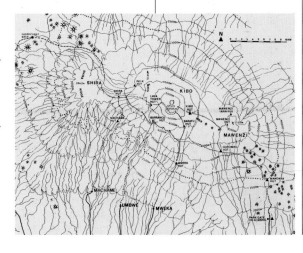

Continental rift

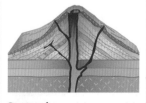

Stratovolcano

Prevalent volcanic activity. Hawaiian, lava flows, lava lakes, fumaroles.

General. Located in the Albert Rift in the Congo, not far from the border with Rwanda, Nyiragongo is part of the Virunga Volcanic Chain. It is a stratovolcano with a very regular cone shape, numerous parasitic cones on the flanks, and a summit crater ¾ miles (1.2 km) in diameter.

Nyiragongo is one of the few volcanoes to have had an active lava lake inside its summit crater; and this one had been active for decades, from 1927 to 1977. On January 10, 1977, following a large fissure-type eruption, the lava lake emptied. The lava flows that followed had an initial speed estimated at 62 miles (100 km) an hour and covered an area of 7¾ square miles (20 km²), killing hundreds of people. Following the rapid emptying of the lava lake, part of the crater structure collapsed, and the mixture of rocks and water came in contact with the magma, generating a hydromagmatic explosion. The cloud of smoke and volcanic ash reached a height of 36,090 feet (11,000 m). The lake began refilling on June 21, 1982, but by October it was already completely solidified. Volcanic activity began again in 1994, but no scientific expeditions were organized to study the zone.

The ethnic war in nearby Rwanda had disturbed the region, bringing into it more than two million refugees who set up camps right at the foot of the volcano. An Italian volcanologist, in the area in 1994–95 for the UN, was able to make at least partial observations of this new volcanic activity. By November 1994 a 165-foot (50-m) high cone had formed inside the crater and was discharging two lava flows that were creating a small lava lake inside the crater.

Between March and April of 1995 the activity of the volcano changed notably with the eruption of large lava fountains. This activity ended at the end of April when the initial cone disappeared, its place taken by a cavity. Then, after a series of earthquakes felt throughout the region, the activity continued in a discontinuous manner with eruptions that lasted only a few hours and followed one another at intervals of several days. In the period between April and August 1995 a volume of lava estimated at 2 billion cubic feet (56 million m^3)—almost three times that of the 1977 eruption—accumulated inside the crater, raising the bottom by about 165 feet (50 m). On January 17, 2002, the volcano erupted, pouring lava into the nearby town of Goma.

AFRICA

Access and principal attractions. *The city of Goma, on the banks of Lake Kivu, is the place to begin the ascent of Nyiragongo. From there you can take a car to Kibati, entrance to the Virunga national park. You'll need to hire bearers and a guide. The ascent takes five hours (the difference in altitude is 5,575 feet [1,700 m]) and winds through a dense forest up to the height of the Shaheru cone. There begins the ascent of the steep slopes of the volcano. After 10 minutes you reach two refuges where you can set up a base camp (you'll need to be well equipped to face the cold nights). The peak can be reached in 20 to 30 minutes of hiking, covering the last 655 feet (200 m) of the climb. At the summit crater, about 1,150 feet (350 m) deep, you can examine the thick layers of lava on its walls crossed by many dikes. Several splatter cones, built up from accumulated strips of fluid lava, stand on its interior. Since 1996 the political climate in the region has made visits risky. Aside from the volcano, Goma has refugees, cholera and various warring bands.*

157

Opposite: The lava lake that had occupied the crater of Nyiragongo since 1927 as it appeared in 1977.
Left: January 10, 1977, the cloud of smoke and volcanic ash rises about 36,090 feet (11,000 m), the result of hydromagmatic activity.

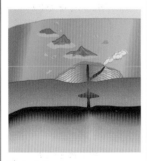

Hot spot

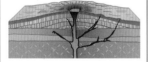

Shield volcano

Prevalent volcanic activity. Hawaiian, lava flows, Strombolian, fumarolic.

General. The volcanic island of Réunion, located in the Indian Ocean 435 miles (700 km) to the east of Madagascar, was created by enormous quantities of magma that rose from the mantle through the moving ocean crust: the magma that accumulated on the ocean floor created a shield volcano with a base diameter of 150 miles (240 km) and about 13,125 feet (4,000 m) in height; of that more than 8,530 feet (2,600 m) are above the surface of the water. The island is elliptical in shape (43½ by 31 miles [70 by 50 km]), 970 square miles (2507 km²), oriented in a northwest-southeast direction. The volcano is composed of two distinct parts. Piton des Neiges (10,070 feet [3,069 m]), located in the northwest area of the island, forms the island's main structure. It is a vast extinct volcanic cone with a base 31 miles (50 km) in diameter and slopes with an average angle of 5–10 degrees. The younger and also very active Piton de la Fournaise (8,632 feet [2,631 m]) forms the southeast portion of the island. It is a cone with a base of 25 miles [40 km] in diameter, and the average angle of its slopes is around 10 degrees. On its summit is the large caldera called Rempart, inside which is a second caldera, called Enclos Fouqué, 5 by 8 miles (8 by 13 km) in size with walls 655 feet (200 m) high. It is a vast horseshoe-shaped depression open toward the sea.

The walls of the two calderas are separated by a plateau of volcanic sand (Plaine des Sables). Inside Enclos Fouqué stands the summit cone, which has two craters, one called Dolomieu, the other called Bory, in honor of the famous French geologist (after whom dolomite is named) and a French naturalist.

Piton de la Fournaise is one of the most active volcanoes in the world. Over the past 50 years it has erupted an average of once every 14 months, and it is known to have erupted 160 times since the 17th century. Most of these eruptions were of the Hawaiian type, meaning they involved lava fountains and flows fluid enough to cover many miles. The final phase of an erup-

tion often involves Strombolian activity. Most of these eruptions have lasted from a few hours to a few months, but there have been cases in which they went on quite a bit longer. Such was the case with the eruption that began in June 1985 only to end in December 1988.

From 1992 to March 1998 the volcano was quiescent; what broke the spell was the opening of a new eruptive fissure on the northern flank of Dolomieu, leading to fountains of lava 165 feet (50 m) high and flows that reached all the way to the sea. The composition of the lava usually varies from tholeiitic basalt to transitional basalt, but during major eruptions the lava erupted is far richer in olivine crystals and is called oceanite. There was further activity in March 2001.

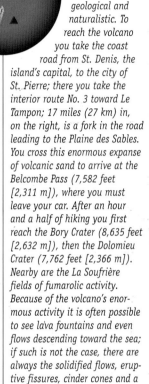

Access and principal attractions. The island of Réunion is of great interest, both geological and naturalistic. To reach the volcano you take the coast road from St. Denis, the island's capital, to the city of St. Pierre; there you take the interior route No. 3 toward Le Tampon; 17 miles (27 km) in, on the right, is a fork in the road leading to the Plaine des Sables. You cross this enormous expanse of volcanic sand to arrive at the Belcombe Pass (7,582 feet [2,311 m]), where you must leave your car. After an hour and a half of hiking you first reach the Bory Crater (8,635 feet [2,632 m]), then the Dolomieu Crater (7,762 feet [2,366 m]). Nearby are the La Soufrière fields of fumarolic activity. Because of the volcano's enormous activity it is often possible to see lava fountains and even flows descending toward the sea; if such is not the case, there are always the solidified flows, eruptive fissures, cinder cones and a great variety of other volcanic structures. It's a good idea to hire one of the authorized local guides; the fog that often envelops the volcano can create serious danger to tourists.

Opposite: Dolomieu Crater, site of most of the volcanic activity of recent years.
Left: One of the many cinder cones around the base and slopes of the volcano.

ASIA AND OCEANIA

The area of the western Pacific and northern Indian oceans is dominated by a major concentration of active volcanoes, most of them grouped in a succession of island arcs. Only a limited number of active volcanoes, about 30 in all, are spread in the northeastern part of the Asian continent, in Korea, China, Mongolia and Russia. Almost all the volcanoes owe their existence to the areas of convergence between the Asian region and the Pacific and Indian plates, moving respectively, northwest and to the north. The speed of approach between the plates varies from $1\frac{1}{2}$ to $4\frac{3}{4}$ inches (4 to 12 cm) a year and the inclination of the plane of descent of the ocean plate into the mantle varies from a few dozen degrees (arc of the Tonga Islands) to the almost vertical (the Marianas). The process of the oceanic plate being drawn into the mantle produces magma rich in water and thus highly explosive. Some of the arcs are islands in the narrowest sense, composed of an uninterrupted chain of volcanic islands; other are composed of large islands (Japan, Java, Sumatra) or peninsula extensions from continental masses (Kamchatka).

Volcanoes of the northern Pacific

Beginning in the northwestern corner of the Pacific there is the large arc composed of the Kamchatka peninsula (70 volcanoes), the Kuril Islands (50 volcanoes) and the volcanoes of the island of Hokkaido (21), the northernmost of Japan's four main islands, beyond which lies the basin of the Okhotsk Sea. At the level of Japan's Honshu island, the system splits into two branches, one moving south and including the Bonin Islands (21 volcanoes) and the Marianas (14 volcanoes); the branch that heads southwest is composed of the island of Honshu (50 volcanoes), Kyushu (8), the Ryukyu Islands and Taiwan (17 volcanoes), continuing south into the Philippines (53 volcanoes).

Volcanoes of the central Pacific

Running across the central Pacific is a series of volcanic islands extending from the islands of New Zealand to the Kermadec Islands, the Tonga Islands and the Philippines, by way of the islands known collectively as Melanesia. From southwest to northwest there are the volcanic arcs of the Vanuatu, the Solomon Islands, New Britain and New Guinea, for a total of 86 active volcanoes.

Volcanoes of the Indian Ocean

The northeast edge of the Indian Ocean is bordered by the large arc of Indonesia, which extends from Sumatra to Java, continuing eastward in a series of smaller islands. Within this Indonesian arc a total of 117 active volcanoes have been identified.

Characteristics of volcanic activity

The eruptions vary from moderately explosive (Strombolian) with emission of lava flows to strongly explosive (Vulcanian and Plinian with the associated pyroclastic

KAMCHATKA
48. Bezymianny
47. Karymsky
KURILS
46. Usu
JAPAN
45. Fuji
43. Unzen
44. Aso
42. Sakura-Jima
IZU
Pacific Ocean
TAIWAN   VOLCANO   Tropic of Cancer
MARIANAS
41. Pinatubo
40. Taal
39. Mayon
PHILIPPINES
PAPUA
NEW GUINEA   Equator
NEW BRITAIN
INDONESIA   SOLOMONS   30. Rabaul
32. Krakatau   31. Lamington
Galunggung
34. Merapi
35. Semeru   SAMOA
36. Bromo   VANUATU
Ijen, Kawah   FIJI
38. Tambora   Tropic   TONGA
of Capricorn
Indian Ocean   AUSTRALIA
KERMADEC
26. White Island
27. Tarawera
28. Tongariro
29. Ruapehu   NEW
ZEALAND

formation of lava domes (Merapi, Unzen, Aso, Usu, Lamington) that periodically explode. Because of the great amount of rain in the monsoon regions and in those areas subject to typhoons, the materials given off in volcanic explosions (ash and pumice) are easily combined with water to form dangerous mud slides (the Indonesian word *lahar* is universally used to indicate such phenomena).

Dangers and benefits of geothermic energy

The high population densities in Asia-Oceania, most of all in those areas with tropical climates, along with the great danger of immediate (explosive eruptions) and secondary (lahars and tsunamis) phenomena, have led to many victims (161,000 in Indonesia alone) in the past. Fortunately, volcanic activity does not have only negative results for humans. The heat brought to the surface by volcanic activity constitutes an energy source that has become important for several countries in the region. In the Philippines, geothermal exploration has been especially fruitful, leading to the identification of six geothermal fields that produce about 2,000 kW-hours, or 20 percent of the country's energy needs.

flows and surges. The most common volcanic structures are central-vent stratovolcanoes with steep slopes and gentle flanks, compound volcanoes and calderas. Some calderas occupy the central part of an ancient cone; others, most of all on large islands (Japan, New Zealand, Java), have varying sizes and were produced by eruptions that emitted hun-

dreds to thousands of cubic miles of magma. Over the course of the last two centuries the Asia-Oceania area has been the scene of large explosive eruptions, such as those of Tambora (1815), Krakatau (1883) and Pinatubo (1991), which discharged large volumes of pumice and ash. Many volcanoes give off slow emissions of lava, leading to the

WHITE ISLAND, NEW ZEALAND

Lat. 37.52 S – Long. 177.18 E – Alt. 1,053 feet (321 m) – 0401-04

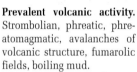

Island arc

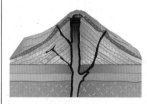

Stratovolcano

Prevalent volcanic activity. Strombolian, phreatic, phreatomagmatic, avalanches of volcanic structure, fumarolic fields, boiling mud.

General. White Island owes its existence to the process of subduction of the Pacific plate beneath that of the Australia-India plate.

The small uninhabited island, 1.9 square miles (5 km²) in all, is composed of a stratovolcano shaped very much like a horseshoe. It lies 25 miles (40 km) off the coast of New Zealand's North Island, in the Bay of Plenty. The volcanic activity is concentrated in the northwestern part of the island.

White Island was seen by James Cook in 1769, but the first observation of its activity dates to 1826. Since then, about 35 eruptions are known to have taken place.

In 1914 an avalanche of the wall of the main crater, with a diameter of 655 to 820 feet

(200 to 250 m), opened toward the south side of the island. The debris buried buildings and killed 11 people working on the island mining sulfur.

Important eruptions took place in 1926–28 (when the Little and Big Donald craters were formed), 1933, 1947, 1955, 1957, 1958, 1959 (with the formation and activity of the Noisy Nellie Crater), 1962, 1965, 1968, 1971, 1976–81, 1984 and, most recently, in 1990–91 and 1992.

Many low-energy seismic shocks were later recorded by New Zealand volcanologists, but the activity seemed to be limited to the feeding of solfa-taras and sulfurous muds until August 1998, when a series of phreatic explosions were observed on the north-western edge of the active crater during the last eruptive

cycle. From October to December 1998 several columns of gas and volcanic ash were erupted from the crater, reaching heights of 0.6 to 1.2 miles (1 to 2 km) above the volcano. There was further activity in the summer of 2000, and since then the civil authorities have been on the alert.

Opposite top: The southeastern coast of White Island. The smoke results from gas emissions from a fissure. Opposite bottom: The island's rocks are altered by the constant activity. Below: The island, wrapped in a plume of smoke produced by fumarolic activity.

Access and principal attractions. *From the town of Whakatane, on the Bay of Plenty, to White Island there are 25 miles (40 km) of sea to cross. You can make the trip in a boat, but when conditions are favorable a helicopter is preferable because you can fly over the entire volcanic complex and, assuming the volcano is in a period of quiescence, you can be put down inside it to make a brief excursion. The view of the broken crater surrounded by the sea is highly evocative. Visible on its inner walls are sulfur deposits and many fumaroles, several of them under high pressure.*

ASIA AND OCEANIA

163

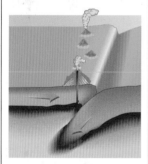

Island arc

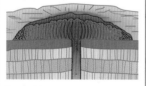

Lava dome

Prevalent volcanic activity. Strombolian, phreatomagmatic, lava domes.

General. The volcano Tarawera on New Zealand's North Island is composed of a complex of domes accumulated above pumice cones. It is located in a region full of lakes, the largest of which, Rotorua, is inside the caldera of the same name. Until the 1886 eruption, the only eruption in historical time, the complex of domes of Tarawera was bordered by Lake Rotomakiriri, with numerous warm springs and a pair of famous silica terraces, the Pink and White Terraces, at the time very popular tourist attractions. Following several earthquakes in the area, the eruption began at 1:30 A.M. on June 10. A fissure suddenly opened along the slope of the northernmost dome, accompanied by numerous explosions. At 2:30 other explosions occurred at the nearby Lake Rotomahana, and by an hour later the fissure had extended 9½ miles (15 km) to the southwest, becoming active along its entire length.

Two hours later the eruption was in large part over, all that remained being several small, localized explosions. One hundred and fifty-three people had died.

The eruption of Tarawera constitutes one of the most interesting examples of how eruptions with contrasting characteristics can occur in the same area. This event, which followed the rapid rise of low-viscosity basaltic magma, produced lava jets from many vents that created alignments of small cinder cones. At the same time, ash and lapilli were shot up to great heights and were dispersed over dozens of miles, forming deposits that have characteristics similar to those typical of Plinian eruptions of more chemically evolved magma. Where the fissure moved into areas of hydrothermal activity, the explosions involved vapor and boiling mud.

Access and principal attractions.

Tarawera is located in a region of lakes and forests that includes some of New Zealand's finest parks. Still quite wild, the area is crossed by only a few roads, most of them open only to off-road vehicles. Route 38 leads from Rotorua to Murupara. From there you can reach the eruptive fissure of Tarawera by climbing from the south along a path about 9½ miles (15 km) long.

Opposite: Aerial view of the large eruptive fracture that opened during the 1886 eruption. It runs across about 9½ miles (15 km) of a mountainous area with many rhyolitic lava domes.

Below: Map showing the entire length of the eruptive fracture; the two colors indicate the area affected by the rain of scoria (gray area) and that most affected by the rain of mud (brown area). The rain of mud occurred after the explosions caused by the magma and the water of Lake Rotomahana.

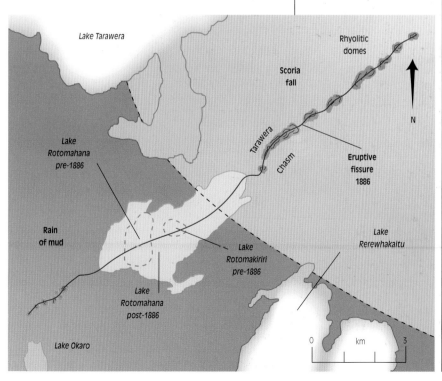

Lake Tarawera

Rhyolitic domes

Scoria fall

N

Tarawera Chasm

Eruptive fissure 1886

Lake Rotomahana pre-1886

Rain of mud

Lake Rerewhakaitu

Lake Rotomakiriri pre-1886

Lake Rotomahana post-1886

Lake Okaro

0   km   3

Island arc

Compound volcano

Prevalent volcanic activity. Strombolian, pyroclastic flows, lava flows.

General. The volcanic complex of Tongariro, located in the natural park of that name on the North Island of New Zealand, is part of the Taupo volcanic system and is, in fact, one of its southernmost elements. Tongariro is one of the most active volcanoes in the park; its activity is estimated to have begun roughly 500,000 years ago. The formation of an ancient elongated massif running northeast was followed by a period of intense glacial erosion that profoundly altered the shape of the original structure. Later new vents opened in different areas of the complex. The North Crater, which began its activity about 50,000 years ago and remained active in postglacial times, has produced numerous lava and tephra flows. At the center of the complex is Red Crater, which has generated numerous eruptions of a'a lava and tephra over the course of the last 10,000 years.

Tongariro, the southernmost volcanic center of the compound structure, is composed of the steep cone of Ngauruhoe, one of the most active volcanoes in the world, which erupts at intervals between two and seven years. Its almost perfectly symmetrical shape has been built up over the last 2,000 years, following a series of explosions, generally of moderate size and Strombolian in nature, associated with the emission of a'a lava flows. The most recent eruption of Ngauruhoe was in 1954, when a series of lava flows with an overall volume of 0.02 cubic miles (0.1 km³) descended the western slopes into the Mangatepopo valley. The emission of magma was accompanied by a rain of ash and lava fountains. During the period January–March 1974 and February 1975 it again began erupting, forming columns of gas and ash that rose up to 5½ miles (9 km) in altitude.

At times the volume of material discharged by the individual explosions was so great that a large part of it fell back down to form avalanches of incandescent material (pyroclastic flows).

The explosive eruptions were accompanied by powerful earthquakes with shock waves perceptible to the human eye as rings of light radiating outward from the crater, while to the ear they were like muffled detonations.

Access and principal attractions. *The Tongariro complex offers visitors the opportunity to see a large variety of volcanic structures as well as natural settings of enormous beauty. An excellent network of paths along with routes for off-road vehicles make it easy to visit all the sites of interest.*

Opposite: The cone of Ngauruhoe seen from the northeast.
Below: Map of the volcanic complex of Tangariro, with the cone of Ngauruhoe, near the bottom, and other craters, including several active in historical time. Access paths are indicated in red.

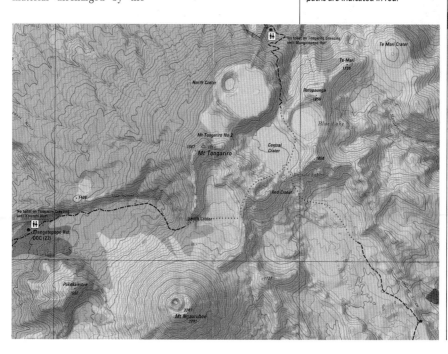

168

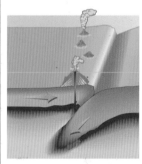

Island arc

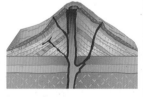

Stratovolcano

Prevalent volcanic activity. Phreatomagmatic, Strombolian, mud flows, fumaroles.

General. The Ruapehu volcano, located on the North Island of New Zealand, is at the southernmost tip of the active volcanic system that begins with White Island and continues southwest across the main island. Thanks to its notable height and the presence of a glacier (Whangaehu), the volcano is a popular area for skiing. The peak is composed of an outer ring 1 mile (1.6 km) long and ½ mile (800 m) wide, inside which is a cone with a central crater occupied by a lake. The water of the lake, normally tepid because of the fumarole at its bottom, drains through a tunnel in the glacier into the Whangaehu valley.

Ruapehu was formed over the course of hundreds of thousands of years (the oldest rocks date to 230,000 years ago), following numerous eruptions, both effusive and explosive. Over the course of the last 1,800 years the volcano has erupted 18 times, producing events with similar characteristics each time. The eruption that began at the end of June 1995 and ended in July of the following year generated about 0.01 cubic miles (0.06 km³) of new material that were dispersed over an area of 11,580 square miles (30,000 km²). Ash from the eruption was deposited more than 186 miles (300 km) away.

In 1995 forceful phreato-magmatic eruptions occurred, produced by the interaction of the lake water with magma. These explosions shot hot water and ash onto the glacier, causing mudflows that swept into the lower river valleys. In 1996 the ongoing evaporation of the water caused by the eruptive activity led to the complete drying of the lake, and the eruptions changed from phreatomagmatic to Strombolian. Further eruptions occurred on September 13, 1999. Despite their modest size, these eruptions caused considerable difficulty, including the closing of 11 airports and the skiing

areas, as well as damage to hydro-electric stations.

In 1953 the sudden destruction of a natural dike that had kept the level of the lake raised more than 26 feet (8 m) caused a flood along the Whangaehua valley; the railroad bridge near Tangiwai was destroyed only minutes before a train on its way from Wellington to Auckland was due to cross. The train plunged into the raging water, taking the lives of 151 people.

Access and principal attractions. *The Ruapehu volcano is located inside Tongariro National Park. It can be reached by a well-equipped network of roads and has areas for skiers (as indicated on the map below) as well as a good number of trails. The peak, the summit glacier and the warm lake in the crater (Crater Lake) are the main attractions of the volcano.*

Opposite: The peak of Ruapehu, with the crater lake and glacier still covered by ash from the September 1995 eruption.
Below: Map of the volcanic area showing the summit glacier, skifields and paths. There are also numerous warnings of lahars.

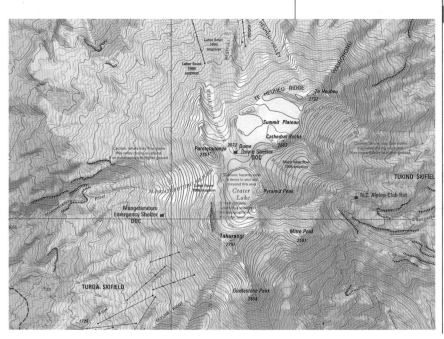

170

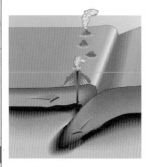

Island arc

Caldera

Prevalent volcanic activity. Phreatomagmatic, Strombolian, Plinian, pyroclastic flows, tsunamis.

General. The caldera of Rabaul, located on the northern end of the Gazelle Peninsula of New Britain Island, forms a deep inlet that has been used since earliest times as a natural port. The city of Rabaul, located on the northern side of the gulf, had a population of 70,000 people until 1994.

The caldera, the result of a collapsed structure, has an overall size of 5½ by 8½ miles (9 by 14 km). It was created following numerous explosive eruptions in prehistoric times. Several volcanic cones surround the caldera, including Kombi, Tovanumbatir and Tuarangunan, while others are inside the caldera itself, such as Tavurvur and Vulcan. The composition of the volcanic products of Rabaul varies from basaltic to rhyolitic.

During historical time Rabaul is known to have experienced major eruptions in 1767, 1791, 1850, 1878, 1937, 1940–43 and 1994. One of the particularities of eruptions at Rabaul is the simultaneous occurrence of volcanic events at both Tavurvur and Vulcan, although they are separated by about 5 miles (8 km). The last crisis at Rabaul began at 6 A.M. on September 18, 1994, when a vent opened on the western side of the crater of the volcano Tavurvur. A little after 7 A.M. activity also began at Vulcan. Over the span of a few minutes the activity at Vulcan had grown in intensity, producing pyroclastic flows that extended roughly 1.2 miles (2 km) into the sea. The high point of the eruptive activity at Vulcan came on the morning of September 19, and on September 24 it declined appreciably. The activity at Tavurvur continued in alternating phases until April 16, 1995. More than 70 percent of the city of Rabaul was completely destroyed by the accumulation of ash and the collapse of roofs. The pyroclastic flow from Vulcan that poured into the sea caused a tsunami that struck the island and drove inland more than 985 feet (300 m). Heavy damage was caused by mudflows and by the corrosion caused by acids in the ash. It has been estimated that the economic losses caused by the eruptions amounted to around $280 million (Australian).

The eruptive episode in 1994 forced authorities to order the evacuation of the city, with the removal of 100,000 people. Over the course of the eruptions only four people were killed. The preventative evacuation of the population was made possible by plans made well in advance by the civil authorities and by the predictions of eruptions made by volcanologists.

Access and principal attractions. Because of the total destruction of the city's infrastructure during the 1994 eruptions, Rabaul can be reached only by way of the new airport at Tokua. The main attraction in the zone is the caldera, occupied by the bay and the volcanic cones of Vulcan and Tavurvur.

Opposite: The 1994 eruption. It began in the Tavurvur Crater on the eastern shore of the caldera.
Below: Map of Rabaul, with its bay created by a series of collapsed calderas. Recent eruptions have taken place simultaneously from and Tavurvur, on opposite sides of the bay.

171

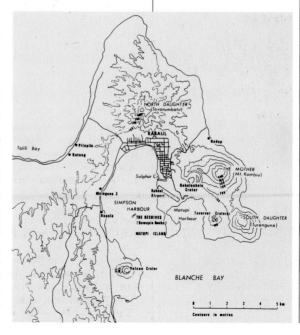

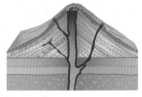

Island arc

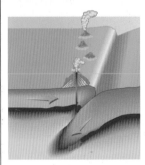

Stratovolcano

Prevalent volcanic activity. Pelean, Vulcanian, lava domes, lahars.

General. Mount Lamington is located 25 miles (40 km) from the northern coast of the Papua Peninsula. It rises gradually from the coastal plain to an altitude of just over 5,500 feet (1,676 m); its peak is essentially a deep horseshoe-shaped summit crater open to the north along Avalanche Valley in the direction of the Ambogo River. The volcanic cone is composed of lava domes and other volcanic deposits left by *nuées ardentes* and lahars.

The volcano, quiescent since prehistoric times and not recognized as a volcano because of its forested peak, suddenly awoke in 1951 with an enormous Pelean eruption that expert volcanologists were able to follow through all its phases, acquiring a great deal of scientific information about these dangerous eruptions.

On January 15 a series of earthquake tremors set off avalanches along the flanks of the volcano; by January 18 the earthquakes had become incessant, and columns of gas and ash were rising from the peak; the next day saw numerous Vulcanian explosions; on January 20 the columns of gas and ash had reached a height of 5 miles (8 km).

At 10:40 in the morning of January 21 a series of more

powerful explosions sent up an enormous volcanic cloud that reached heights up to 9½ miles (15 km); a further violent eruption came that evening.

The pyroclastic flows that followed descended all around the volcanic cone, devastating an area of 91 square miles (235 km^2) and destroying all inhabited centers within 6 miles (10 km) of the volcano. The eruption caused the death of 2,942 people.

Most of the more serious damage was done to the area to the north, where there were no obstacles to block the pyroclastic flows. There were then periods of quiet alternating with periods of intense volcanic activity (on January 25, February 6, February 18, March 5). The highly explosive phase lasted only two months, and in June Mount Lamington entered a new phase of activity. For six years, until 1956, there were extrusions of lava domes. During the initial phase, in the days immediately following the eruption, a lava dome formed on the crater. From February 3 to 9 it rose 100 feet (30 m) per day; in six weeks it had reached a height of 1,475 feet (450 m) from the floor of the crater.

On May 5 this dome partially exploded in a large Pelean eruption; after that the volcano went back to extruding viscous lava, again forming a dome, this one even higher thanks to the formation, on its summit, of a solid spine of lava. The single spine reached a height of 395 feet (120 m), giving the entire structure a volume of 0.2 cubic miles (1 km^3) and the respectable height of 1,772 feet (540 m) from the crater floor.

Opposite: The shape of the dome in 1951, covered by tropical growth. Below: Map of the volcanic area of Mount Lamington. The broken line indicates the boundary of the area partially destroyed during the eruption of 1951; the area inside the solid line was totally destroyed.

Access and principal attractions. *From Port Moresby you must cross the peninsula to the city of Popondetta. There you can organize an excursion to climb Mount Lamington. The main attractions are the horseshoe-shaped summit crater, with its large dome topped by a spine of solid lava, the deposits from pyroclastic flows and lahars, Avalanche Valley and the equatorial forest surrounding the volcano.*

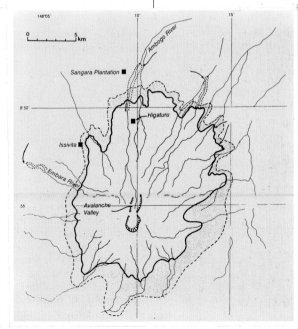

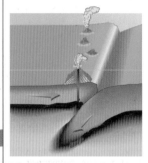

Island arc

Caldera

Prevalent volcanic activity. Plinian, pyroclastic flows, Strombolian, lava flows, tsunamis.

General. Krakatau (also spelled Krakatoa or Krakatao and known in Malay as Rakata) is the name of a volcanic island located in the center of the Sunda Strait between Java and Sumatra. Before its 1883 eruption, the island had been composed of a row of three volcanic cones inside a prehistoric caldera. On May 20 of that year one of these cones, Perbuwatan, located at the center of the prehistoric caldera, began activity. After three months of lesser explosive activity the nearby cone of Danan also began to erupt. On August 26 violent eruptions began, reaching their high point on August 27. The brief span of these events makes them even more astonishing: Krakatau discharged 5 cubic

174

miles (20 km³) of volcanic pumice and ash, and the rapid emptying of the magma chamber led to the collapse of the structure, forming a circular caldera 4 miles (6 km) in diameter and 0.6 miles (1 km) deep. The eruption also produced violent seaquakes that resulted in 100-foot (30-m) high tsunamis in the strait; these destroyed cities and towns all along the coast, killing more than 36,000 people.

The effects of this cataclysm were enormous; the sound of the explosion was heard as far off as Australia, 2,985 miles (4,800 km) away, and the shock wave, felt throughout the world, went around the entire circumference of the planet three times, making a complete trip every 36 hours. Volcanic dust rose 6 miles (10 km) into the atmosphere and spread to the southwest, wrapping around the Earth at the height of the tropical zone and encircling the equator in just two weeks. Because of this dust in the atmosphere the temperature of the Earth was lowered for three years following the eruption (in Europe, solar radiation went down 10 percent). The pumice erupted into the sea covered thousands of square miles, and floating pumice clogged the Sunda Strait. Undersea banks were formed, some several feet thick. The great eruption was followed by a period of quiet that lasted 44 years. In 1927 volcanic activity began inside the caldera. The accumulation of volcanic material led to the formation and appearance above sea level in 1952 of a small cinder cone named Anak Krakatau (Child of Krakatau). This cone grew and produced frequent Strombolian eruptions and emissions of lava flows.

Access and principal attractions. A visit to Krakatau begins in Jakarta, where you can make reservations with one of the many tourist agencies to take an excursion to the volcano by boat. Products of the 1883 eruption are visible along the coasts of the islands of Sertung (Verlaten), Panjang (Lang) and Rakata. Anak Krakatau produces frequent Strombolian-type eruptions accompanied by the emission of lava flows.

Opposite: Explosion atop Anak Krakatau, born in 1927 and visible above water level since 1952.
Below: Maps of the islands in the Krakatau group, before the eruption in 1883 (to the left) and after it. The dotted lines indicate bathymetric depths: the outer curve indicates a depth of 820 feet (250 m), those on the inside are 885 feet (270 m) deep.

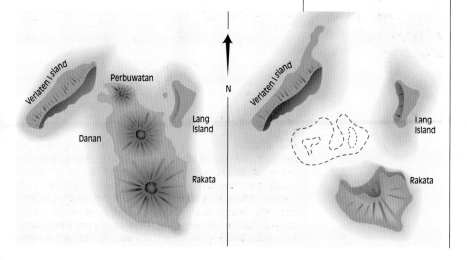

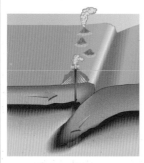

Island arc

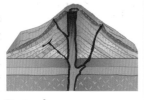

Stratovolcano

Prevalent volcanic activity. Strombolian, phreatomagmatic, pyroclastic flows, lava domes, lava flows, lahars, fumaroles.

General. Galunggung lies about 62 miles (100 km) southeast of the city of Bandung in the western end of the island of Java. The central part of the volcanic cone is occupied by a large horseshoe-shaped caldera, open to the southeast, that was produced following a gigantic avalanche several thousand years ago. The avalanche decapitated the cone and dumped a mass of debris onto the plain below, covering a surface area of 67 square miles (175 km²) and forming a series of hummocks, small hills 33 to 263 feet (10 to 80 m) in height, in the area known as the Ten Thousand Hills of Tasikmalaya, located around the city of that name.

During recent centuries Galunggung has erupted in 1822, 1894, 1918, 1982–83 and 1984. Most of these eruptions ended after several hours or days. For example, the eruption of 1822 began on October 8 with a brief explosive phase that led to pyroclastic flows and lahars that caused the deaths of about 4,000 people. The eruption continued, the extruded lava forming a dome, and by the first of December of that same year it had ended. The eruption of

1982–83 was different. It was preceded by the evacuation of more than 60,000 people; the event itself was divided into three different phases. During the first phase (April–May 1982) pyroclastic flows and lahars descended the southeast slope along the channel left by the ancient avalanche. The second phase (May–October) involved strong phreatomagmatic explosions, caused by contact between the magma and subterranean water, in the course of which clouds of ash and gas rose to between 6 and 12 miles (10 and 20 km) in the air. The close of this phase saw the creation at the summit of the volcano of the circular Walirang crater, 2,625 feet (800 m) in diameter and 495 feet (150 m) deep, which is today considered a maar. During the third and final phase (November 1982 to January 1983) there were emissions of lava flows accompanied by Strombolian explosive activity. The change of eruptive "style" was accompanied by changes in the composition of the magma, which went from andesitic to basaltic.

During the course of the second eruptive phase, three commercial airplanes crossing the zone of the volcano suffered damage (without serious consequences) caused by the abundance of volcanic ash in the atmosphere. These incidents brought to the attention of the world, for the first time, the question of the safety of air traffic during explosive volcanic eruptions.

Access and principal attractions. *Galunggung can be reached by car from Jakarta. The route takes you to the southeast, past the city of Bandung, to Tasikmalaya, at the foot of the volcano. The so-called Ten Thousand Hills in the area are a spectacular example of hummocks, typical deposits left by debris avalanches. The horseshoe-shaped depression at the center of the volcano, the structure left by the breach that caused the avalanche, is the site of many fumaroles.*

Opposite: The Walirang maar inside the summit crater. A great deal of fumarolic activity takes place along the edges of the maar and from near its center, site of a submerged cinder cone produced during the last phase of the 1982–83 eruption. Left: One of the clouds of smoke, ash and volcanic gas produced during the phreatomagmatic phase of the 1982–83 eruption. These clouds reached altitudes of 6 to 12 miles (10 to 20 km), creating serious problems for air traffic.

Island arc

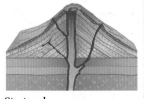

Stratovolcano

Prevalent volcanic activity. Pelean, lava flows and domes, lahars.

General. Merapi, located at the center of the island of Java, is one of the most active and dangerous volcanoes in Indonesia. Composed of two volcanic edifices, only one of which is still active, it dominates the major city of Yogyakarta, with its population of more than one million. Only 19 miles (30 km) separate the city from the volcano. Because of the high viscosity of the magma, andesitic and very high in crystals, the lava discharged by Merapi has difficulty flowing down the slopes of the volcano and tends to form lava domes. These domes slowly grow larger and larger, becoming subject to frequent avalanches due to their instability. Their collapse results in avalanches of lava fragments and incandescent ash and cinders (glowing pyroclastic flows of the Pelean or *nuée ardente* type) that sweep quickly down the channels cut in the volcanic cone. They represent a great danger to the villages located at the base of the volcano. In some cases the avalanches cause the rapid depressurization of sections of still hot magma, leading to small explosions on the sides of the dome.

Geological studies indicate that this type of activity has been going on here continuously since the sixth century, with an alternation of explosive and effusive phases making continual changes to the topography and altitude of the volcano. In 1672 a pyroclastic flow killed about 3,000 people. From the end of the 1800s until now, the six years between 1924 and 1930 were the volcano's longest period of inactivity. At the end of that brief respite, in 1930, the volcano returned to violent activity with an eruption that killed 1,400 people. From 1972 to the early 1990s, the volcano went through a period of practically uninterrupted activity, with the continuous accumulation of lava on its summit and periodic avalanches with pyroclastic flows. It has erupted as recently as March 2001.

On November 22, 1994, an avalanche of incandescent blocks surprised the inhabitants of the villages on the southern slope while they were at work. Sixty-three were killed, and another 500 suffered burns. In October of the next year another avalanche left 41 dead and 280 wounded. Merapi presents a potential danger not only because of its Pelean activity but because of the danger of lahars, which are produced by the rains of the monsoon season and which can sweep away the loose material accumulated on the volcano's slopes.

Access and principal attractions. Various paths lead to the peak of Merapi. The simplest and most interesting leaves from the mountain town of Selo, located at 5,250 feet (1,600 m) of altitude in the saddle between Merapi and the Merabu volcano (10,309 feet [3,142 m]). The trip up and back to Selo takes seven to eight hours. The path up is relatively easy to follow and does not present great difficulties in terms of mountain climbing. Given the danger of Pelean activity you should inquire about the current condition of the volcano, and you must obtain permission to climb to the peak; for this you must go to the local police station and make use of an experienced guide. The visit to the peak, the lava dome and the frequent avalanches of incandescent material are the volcano's main attractions.

Opposite: The cone of Merapi. The deep gorge visible in the photograph was created by the many pyroclastic flows sweeping away any loose material on the slope of the cone. Most recent avalanches have involved the material from collapsed summit domes flowing down this channel.
Left: Aerial view of the peak of Merapi, with a lava dome in formation.

Island arc

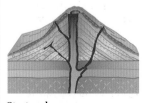

Stratovolcano

Prevalent volcanic activity. Pelean, Strombolian, lava flows and domes, lahars.

General. Located in the eastern region of Java, Semeru is the highest volcano on the island as well as one of its most active and dangerous. In fact, the history of this volcano includes roughly 70 eruptions since 1818 in an area occupied by many cities: to the east Lumajang, to the south and southeast Pronojiwo, Tempeh, Pasirian and Candipuro.

Since 1946 Semeru's activity has consisted primarily in the formation of lava domes inside its summit crater, known as Djonggring Seloko, 1,640 by 2,130 feet (500 by 650 m) in size and about 720 feet (220 m) deep. These domes have often resulted in avalanches of incandescent material (Pelean activity). Semeru has been in almost continuous eruption since 1967 with the extrusion of a lava dome; frequent avalanches alternate with lava flows and jets of cinders. In 1992 a flow of basaltic lava about 1 mile (1.5 km) long with a volume of 53 million cubic feet (1.5 million m³) covered the valley of Besuk Kembar.

During recent years the activity of Semeru has included frequent Strombolian eruptions, with explosions in the summit crater occurring roughly every 30 minutes. Numerous, more intense explosions have also been registered, probably of the Vulcanian type, shooting gas and ash several miles into the atmosphere.

Since March 1997 airline companies have announced repeated warnings of ash in the air over the zone at altitudes between 5 and 6 miles (8 and 10 km). As of yet, fortunately, there have been no accidents.

Access and principal attractions. *Semeru is in the eastern section of the island of Java. Malang, about 25 miles (40 km) away, is the best city from which to begin a trip up the volcano. Given the possibility of Strombolian and Pelean activity, it is best to obtain up-to-date information on conditions; you will also need to visit a police station to obtain the proper permit to climb the mountain. You will need the proper equipment, and a guide is highly recommended.*

Leaving Malang you pass the village of Tupang to reach the small town of Rano Pani (7,220 feet [2,200 m]) and its lake. If you don't already have one, you can hire a guide here; you may also need porters. This is where the ascent of Semeru begins. About 12 hours are needed to reach the peak. A 6-mile (10-km) long path winds its way from the town in a southwesterly direction to Lake Rano Kukmbolo (7,875 feet [2,400 m]). From there another path leads 17½ miles (28 km) through dense jungle to the base camp of Arcopodo. The next morning the ascent of the peak begins. The final stretch is the most difficult, the ground being covered by ash and scoria. The ascent of Semeru, the highest mountain in Java, with its almost perfect conical shape, is considered a "classic" of mountain climbing. Along the way up there are many deposits from pyroclastic flows and lahars to examine as well as Strombolian activity to witness, with incandescent material shot into the air.

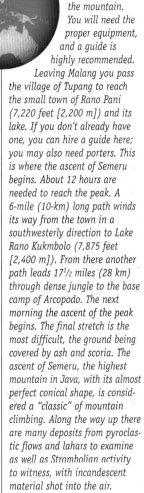

Opposite: Strombolian eruption from the summit of the volcano.
Left: Large explosion from the central crater of Semeru.

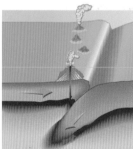

Island arc

Cinder cone

Prevalent volcanic activity. Strombolian, fumarolic.

General. Bromo owes its name to the Hindu divinity Brahma. It is composed of a cinder cone located, together with other eruptive centers not active in historical time (Batok, Widodaren, Kursi), inside the caldera of Tenegger. This caldera is part of the Bromo-Semeru volcanic massif at the eastern edge of the island of Java. With its 5¼ by

6 miles (8.5 by 10 km) of extension and 360-foot (110-m) depth, it ranks among the world's largest calderas. The cone of Bromo has an elliptical crater 2,625 feet (800 m) long in the north-south direction and 1,970 feet (600 m) running east-west. The volcano's eruptive history dates back to the 17th century and boasts almost 60 eruptive periods since 1767. The eruptions are usually Strombolian, with jets of cinders, lapilli and lava bombs that fall back on to the cone, contributing to its growth. Less frequent are episodes of Vulcanian activity, tied to the interaction of magma with subterranean deposits of rainwater, and phreatic explosions. Large eruptions took place in 1980, 1989, 1995 and 2000. Explosions in June 1980 threw incandescent bombs 3 feet (1 m) in diameter up to 1,315 feet (400 m) from the rim of the crater, a spot often visited by tourists. The 1995 eruption gave off clouds of gas and ash that rose up to 2,300 feet (700 m). The eruption of March 3, 1995, caused a light rain of ash to the south and southeast, reaching up to 12½ miles (20 km) from the volcano.

Access and principal attractions. *Leaving from Ngadisari, located 2 miles (3 km) below the large caldera, you reach Cemoro Lawang (6,575 feet [2,004 m]), site of a hotel and guest house.* From there the base of the volcano, 2 miles (3 km) away, can be reached in a one-hour walk. The route, which can be done on foot or horseback, is truly spectacular. Volcanic ash forms a true "sea of sand" that moves into immense fields of lava, quite often immersed in fog. The ascent to the crater is simple but also somewhat tiring. The large crater of Tengger with its seven volcanoes, the spectacular view from the peak of Bromo, the expanses of volcanic sand and the villages are all elements of great interest.

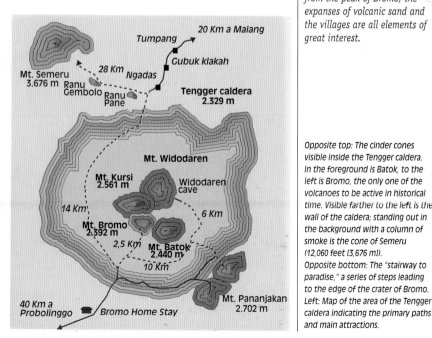

Opposite top: The cinder cones visible inside the Tengger caldera. In the foreground is Batok, to the left is Bromo, the only one of the volcanoes to be active in historical time. Visible farther to the left is the wall of the caldera; standing out in the background with a column of smoke is the cone of Semeru (12,060 feet [3,676 m]).
Opposite bottom: The "stairway to paradise," a series of steps leading to the edge of the crater of Bromo.
Left: Map of the area of the Tengger caldera indicating the primary paths and main attractions.

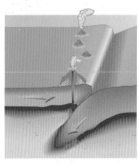

Island arc

Stratovolcano

Prevalent volcanic activity. Phreatic, lahars, fumaroles.

General. Kawah-Ijen is located in the easternmost area of Java, in the volcanic massif of Ijen, within an enormous caldera 10 by 12½ miles (16 by 20 km) that hosts another active volcano, Raung. Kawah-Ijen (the name means "green crater," from the color of the water in its lake) has a surface area of 0.15 square miles (0.4 km²) and walls 655 feet (200 m) high.

The lake is surrounded by a great number of high-temperature fumaroles that deposit sulfur on the rocks. Large deposits of malachite are visible along the flanks of the crater. The sulfur is mined, but in a cruel and dangerous way: by hand, with laborers who must carry baskets loaded with 176 to 187 pounds (80 to 85 kg) of sulfur from the crater floor up along a narrow path to the small village of Licin, 16 miles (25 km) away, where it is purchased and worked by a state-owned industry. In these inhuman conditions the average life span is 30 to 40 years.

The lake water contains a great quantity of dissolved volcanic gas, so much that the water is extremely acidic. In fact, the 1.34 billion cubic feet (38 million m³) of lake water are estimated to contain 1.4 million tons of chloric acid, 600,000 tons of hydrochloric acid and 550,000 tons of sulfuric acid, at an average temperature of 147°F (64°C).

In 1796 a group of Dutch scientists observed the volcano's first recorded eruption. This was followed by others in 1817, 1917, 1936, 1952, 1993 and 2001.

Such eruptions, primarily explosive, make the acidic waters of the lake overflow, causing great damage to the surrounding environment. During the period of Dutch colonial rule, efforts were made, including a spillway and dike constructed on the western wall of the volcano, to control the level of water and diminish it if an eruption appeared imminent. The great quantity of volcanic gas that mixes into the lake water periodically changes the color of the water, from green to turquoise to gray. The gas, full of hydrochloric acid and sulfur dioxide and therefore extremely poisonous, sometimes collects in large bubbles that burst on the surface, transforming the crater into a true death trap. In 1976 an event of this kind killed 49 sulfur laborers. Another 25 died in 1989.

Access and principal attractions. *A trip to Kawah-Ijen is no simple undertaking. Leaving from Jakarta means having to cross the entire island of Java from west to east to reach the city of Banyuwangi. From there you travel to Licin, the village with the path leading to the volcano. If you depart instead from the city of Bali, you need only*

cross the strait and travel to Licin. The major attractions include the amphitheater of the crater of Kawah-Ijen, with its greenish lake and enormous quantities of deposits. There is the ongoing work of extracting and transporting the mineral, along with the outcroppings of malachite on the internal walls of the crater. Be very careful: the water gives off toxic fumes without warning.

185

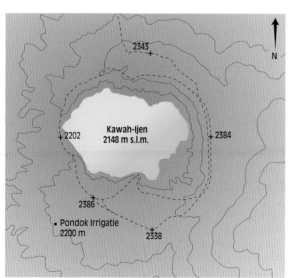

Opposite: The lake inside the summit crater.
Above: Sulfur gatherer at work.
Left: Topographical map of the summit area of the volcano with the path that leads to Licin.

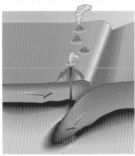

Island arc

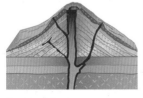

Stratovolcano

Prevalent volcanic activity. Plinian, pyroclastic flows, lahars, fumaroles.

General. Tambora is located on the island of Sumbawa in a position that puts it out of line with respect to the arrangement of Indonesia's main chain of volcanoes. As a result of its location, the magma that feeds Tambora is richer in potassium than the typically calc-alkaline magmas of most Indonesian volcanoes, its composition coming closer to the alkaline-potassic series. Tambora is a large stratovolcano with numerous parasitic cones covering a surface area of about 540 square miles (1,400 km^2); the central part is occupied by a circular caldera 4 miles (6 km) in diameter with a depth between 1,970 and 2,295 feet (600 and 700 m), formed following the enormous eruption in 1815. This event, which erupted a volume of magma estimated at 12 cubic miles (50 km^3), is commonly considered the largest volcanic eruption in historical time. The eruption directly or indirectly caused the death of 92,000 people and had enormous effects on the world's climate for several years. Global temperatures were lowered by as much as 5°F (3°C), and a year after the eruption the amount of volcanic material in the atmosphere absorbed so much solar radiation that in parts of

Europe and North America 1816 was known as the "year without a summer."

The eruption followed a period of volcanic inactivity that had lasted since the eighth century. Thus the violence of the eruption was made even more astonishing by coming from a volcano that until then seemed to do nothing but emit lava flows. Tambora did not suddenly "awaken," however, and signs of activity had begun at least a year earlier. On April 10, 1815, it entered a phase of paroxysmal eruptions that lasted more or less 24 hours, during which a Plinian column rose 19 miles (30 km) in altitude. After three hours the Plinian phase ended, followed by a period of pyroclastic flows that devastated the slopes of the volcano, pouring an enormous quantity of gas, ash and pumice into the sea and across the surrounding land. The caldera then collapsed, and with it the cone, cutting the mountain down by as much as one-third of its height. The eruption killed 10,000 people outright on the island of Sumbawa, but an even larger number (around 82,000) perished later in the terrible famines brought on by the destruction and failure of crops on the islands of Sumbawa and Lombok.

Enormous quantities of dust and gas from the eruption rose into the stratosphere and were dispersed throughout the world, lowering temperatures and causing colder seasons on a global scale. The general lowering of temperatures had adverse effects on agriculture in many countries, leading to famines accompanied by epidemics of diseases, thus adding more victims to the total.

Access and principal attractions. From the city of Dompu a road for off-road vehicles leads along the coast of the Tambora Peninsula. There are several villages within a day's walk of the peak of the volcano. The view from the summit is spectacular, taking in the entire caldera and the coastline of Saleh Bay.

Opposite: The caldera of Tambora, 4 miles (6 km) in diameter with walls up to 2,295 feet (700 m) high. It came into being in 1815 following the collapse of a large part of the volcano. Sulfurous gases rise from the volcano, and the ground is covered with white ash.
Below: Map showing the extent of ashfall from the 1815 eruption. The values are given in inches (1 inch = 2.54 cm).

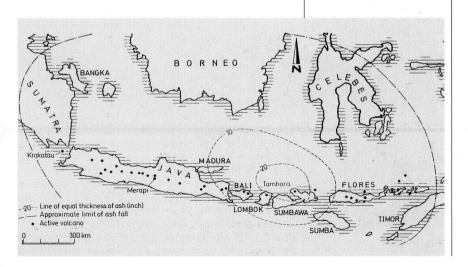

Line of equal thickness of ash (inch)
Approximate limit of ash fall
• Active volcano

0   300 km

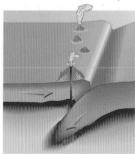

Island arc

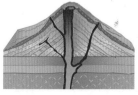

Stratovolcano

Prevalent volcanic activity. Strombolian, Vulcanian, pyroclastic flows, lava flows, mudflows.

General. Mayon, located in the southeastern peninsula of the island of Luzon, about 200 miles (325 km) southeast of Manila, is a classic and spectacular example of a stratovolcano with a symmetrical cone. The cone has a small summit crater about 655 feet (200 m) in diameter, and its base is surrounded by deposits from lahars, the result of volcanic ash being swept down the slopes by rains, which are quite abundant here. Mayon is one of the most active volcanoes in the world. Since the 17th century there have

been at least 45 eruptions, the most recent of which dates to July 2001. The eruptions of 1814 and 1875 caused, respectively, the deaths of 1,200 and 1,500 people, mostly as a result of lahars set off by monsoon rains. In 1897, 350 people died following pyroclastic flows that devastated the villages of Legaspi, Daraga and Santo Domingo. The major eruptions of the 1900s occurred in 1928, 1968 and 1984. During the paroxysmal phase of the 1968 eruption columns of ash and gas were generated that reached heights around 6 miles (10 km), accompanied by numerous pyroclastic flows down the slopes of the cone. The final phase of the eruption involved the emission of a'a-type lava flows that descended the southwestern slope over a distance of 11,485 feet (3,500 m).

An explosion during the 1984 eruption destroyed the southeastern side of the crater. Most of the pyroclastic flows during the next days came from that opening, digging a deep gorge in the slope about 820 feet (250 m) deep and 820 feet (250 m) wide extending downhill about 2¼ miles (3.5 km).

On February 2, 1993, an eruption began suddenly with a phreatomagmatic explosion that sent a pyroclastic flow down the southeastern slope. This flow, 4 miles (6 km) in length, killed 75 farmers and injured more than 100. Today, a modern system of instrumental monitoring is installed on Mayon, including seismic stations and instruments for registering ground movement. These are divided among three different observatories: one in Santo Domingo, 4⅓ miles (7 km) east of the crater; one in the Mayon Rest House, 4⅓ miles (7 km) to the northeast of the volcano at 2,495 feet (760 m) altitude; and one at Lignon Hill, located in the Legaspi airport 7½ miles (12 km) from the crater.

Access and principal attractions. *Only one path leads up the volcano, and it begins from the north, at the Mayon Rest House. The path follows a series of cuts made in an old flow of gray lava. A hike of three to four hours brings you to the end of the vegetation, and at that point (6,070 feet [1,850 m]) it is possible to set up camp. Another two to three hours' hike, clambering across highly unstable terrain with the constant risk of falling rocks, reaches the peak of the volcano. It must be kept very much in mind that following the eruptions of 1993 and 2001 the ascent of the volcano is not advisable, for there is always the possibility of another sudden eruption. Mayon is famous throughout the world for the almost perfect symmetry of its cone. The volcanic deposits along its slopes consist of fields of a'a lava that can be crossed only with extreme difficulty because of their jagged surface, pyroclastic flows with breadcrust bombs and layers left by lahars, including blocks of material of various kinds soldered together by ash.*

Opposite: The cone of the volcano, seen from the southeast. The deep gorge visible in the photo was created during the 1984 eruption.
Left: Pyroclastic flow descending the slope of Mayon.

TAAL, PHILIPPINES

Lat. 14.00 N – Long. 120.99 E – Alt. 1,312 feet (400 m) – 0703-07

Island arc

Stratovolcano

Prevalent volcanic activity. Phreatomagmatic, Strombolian, lava flows, fumaroles.

General. The volcanic edifice of Taal, located in the Batangas province about 37 miles (60 km) south of Manila, is composed of a 15½ by 18½ mile (25 by 30 km) caldera occupied by a lake, Lake Taal, inside which is a 3 by 3¾ mile (5 by 6 km) island. This island, known as Volcano Island, has been the location of all historical eruptive activity. The island itself has its own small crater lake, about 230 feet (70 m) deep, along with numerous other craters and volcanic cones.

The eruptions of Taal have been among the most destructive and violent of the Philippine archipelago. The first of which there is historical record occurred in 1572. Since then there have been more than 30 others, the most recent in 1999, with intervals of rest periods, the longest of which lasted 62 years. Most of these eruptions have been Strombolian, accompanied by the emission of lava flows. In some cases lake water penetrated the eruptive vent, rapidly vaporizing in contact with the magma and producing violent hydromagmatic explosions. There was minor phreatic activity in 1976 and 1977.

One of the major historical eruptions was that of 1749,

which occurred in a fracture running northeast-southwest across the central part of the island; over the course of the event several other vents opened, including one where the crater lake stands. The surrounding area was covered by ash and lapilli. In 1754 there was an even larger eruption. On May 6, the volcano began giving off smoke clouds, which in the days to come formed a column several miles high from which lapilli and ash fell. The activity continued at this level for the following months, accompanied by powerful earthquakes. The fall of volcanic material destroyed crops within a 6-mile (10-km) radius of Lake Taal, and the amount of material in the air put the area in total darkness for many days. On November 25 the level of the lake rose suddenly, flooding villages as far as 4⅓ to 5 miles (7 to 8 km) away. The eruption ended in December of that year, after having destroyed numerous villages and killed an unknown number of people.

The eruption of September 28–30, 1965, was hydromagmatic, following the interaction of lake water with magma. The explosions produced columns of smoke and ash 10 to 12½ miles (16 to 20 km) high that picked up lapilli and ash suspended in the gas and water vapor and formed ring-shaped clouds that swept from the base of the column at great speed, destroying anything in their path.

These events led to that particular type of eruptive phenomena being recognized and described for the first time. On the basis of the phenomena's similarity to the ring-shaped clouds that sweep radially around the base of the mushroom clouds of atomic explosions, it has come to be called a base surge.

Opposite: The crater lake inside Volcano Island. The small island is the center of historical activity in the caldera of Taal.
Below: Map of the volcanic area of Lake Taal (Lake Bombon).

Access and principal attractions. *Lake Taal is an hour and a half from Manila by way of the city of Tagatay and heading in the direction of the Taal Monument. The departure point for a visit to the volcano is the village of Talisay on the north bank of the lake. You must rent a motor boat to reach Volcano Island; from there a 40-minute walk gets you to the belvedere on the small crater lake. To visit the eruptive sites of Tabaro, the center of volcanic activity since 1965, you must moor on the western side of the island, from where you reach the crater in an hour and a half by foot.*

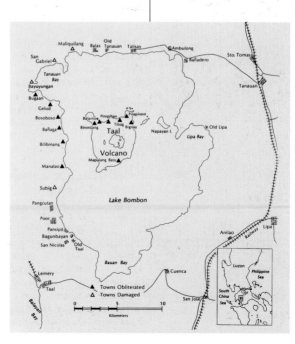

192

Island arc

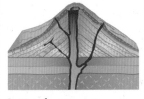

Stratovolcano

Prevalent volcanic activity. Plinian, pyroclastic flows, lava domes, mudflows.

General. Pinatubo is located on the borders of three provinces, Pampanga, Zambales and Tarlac, in the central area of the island of Luzon. It is part of the western volcano chain that extends more than 137 miles (220 km) in a north-south direction. The volcano, which began its activity around 1.1 million years ago, is composed of a collection of lava domes surrounded by thick layers of pyroclastic material. The volcano's history involves eruptive periods separated by many centuries of quiescence during which the dense tropical vegetation covers the entire volcano. The eruptions are primarily explosive, large in size, involving the emission of many cubic miles of ash and pumice, often followed by the formation of lava domes.

In June 1991, after nearly 400 years of inactivity, Pinatubo produced one of the largest explosive eruptions of the 20th century. The first indications of the volcano's reawakening came early in April, with the formation of a fissure along the northwestern slope of the volcano that gave off ash and gas. On April 5, a seismic station located to the northeast of the volcano registered 223 earthquakes, an indication that the volcano was undergoing intense fracturing. A series of minor explosive events began on June 12, ending on June 15, when the paroxysmal eruption began. At 10:27 that morning a Plinian column of gas, ash and pumice rose to a height of 25 miles (40 km). The column's giant umbrella spread over a few hundred miles, blacking out the sky and raining pumice and ash on an area of more than 115,000 square miles (300,000 km²). A few hours later the volcano produced a series of massive pyroclastic flows that devastated the valleys near the volcano and up to a distance of roughly 10 miles (15 km). The final phase of the eruption involved the destruction of the summit of the volcano with the formation of a caldera 1.2 miles (2 km) in diameter. In the destruction of the summit about 985 feet (300 m) was removed from the mountain. The volcanic activity subsided over the coming days, limiting itself to the simple emission of plumes of ash and gas.

Thanks to the collaboration of Philippine volcanologists and members of the United States Geological Service (USGS), a series of studies of the eruptive behavior of the volcano had been carried out in the months before the paroxysmal eruptions began, and a network of monitoring stations had been installed that enabled the scientists to recognize and interpret the signs given off by the volcano. On the basis of the information collected, the risk of eruption was recognized in time to evacuate part of the resident population, in total about 1.8 million people. Given the size of the eruption and the enormous dangers it created, the direct consequences on the population were small: the eruptions killed 883 people, with an additional 184 injured and 23 missing. Most of these resulted from the mudflows that occurred in the months after the eruption. The economic losses, however, were enormous: 651,000 people were

Opposite: The caldera of Pinatubo seen from the east. It came into being following the 1991 eruption and is occupied by a lake. The two islands are the top of a lava dome that formed during the months after the eruption. The patch of white in the lake to the right is caused by warm water rising from the bottom of the crater.
Below and bottom: Effects of the eruption on nearby inhabited areas. The church of Bacalor near San Fernando, partially buried by lahar deposits; a collapsed hangar at Clark Air Base, 12½ miles (20 km) to the east of the caldera.

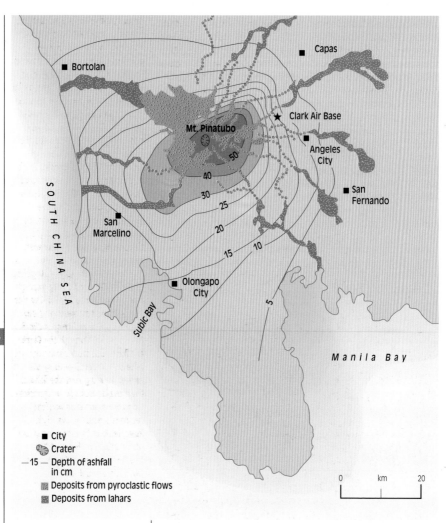

City ■

Crater 🕸

Depth of ashfall in cm —15—

Deposits from pyroclastic flows ▦

Deposits from lahars ▦

0   km   20

Above: Map of the volcanic deposits from the 1991 eruption.

put out of work, and Clark Air Base, one of the largest in the world, famous as a base of operations during the Vietnam War, was rendered unusable and had to be abandoned.

One of the most dramatic results of the eruption was the massive movement of ash and

pumice caused by typhoons in later years. Of the estimated 1.7 to 2.4 cubic miles (7 to 10 km³) of material accumulated on the slopes of the volcano, 70 percent has been swept away and carried downhill in the form of immense mudflows (lahars).

These mudflows buried towns, knocked down power lines, swept away roads and caused more victims than the eruption. In efforts to limit the damage caused by lahars, the Philippine authorities have undertaken large-scale engineering works aimed at channeling the rivers of mud. Over the months and years following the eruption, artificial embankments have been built to prevent the flooding of inhabited and cultivated areas.

Decisive human efforts undertaken before, during and after the 1991 eruption reduced the loss of human life and mitigated the financial losses. Even so, the eruption caused an enormous disturbance in the geological balance of a vast area, a disturbance that will have effects on the inhabitants and the environment for many years to come.

Access and principal attractions.
Barring unforeseen changes in the topography of the upper areas of the volcano, the kind of changes usually caused by the erosion of pyroclastic deposits by rain, the caldera can be reached on foot by ascending the north slope of O'Donnell Valley. The summit of the volcano with the caldera lake offers a desolate and fascinating spectacle. Thick layers of pyroclastic debris and deposits from lahars are visible along the route and in the valleys around the volcano, in particular along the valley of the Sacobia River, on the east side of the volcano, easily reached from the city of Angeles, 15 miles (25 km) to the east of the summit crater. From Angeles you will need permission from the Philippine military authorities to travel through the Clark Air Base, but from there you can hike over several miles of the valley, viewing first the lahars, then the deposits from pyroclastic flows exposed in vertical sections several yards high. Vast fields of lahar deposits are also visible on the road that rises toward Lake Mapanuepe from San Marcelino.

Left: Column of gas and ash, 12½ miles (20 km) high, rising over Pinatubo on the morning of June 12, 1991. Similar columns rose on June 13 and 14, and following the major eruption on June 15 a Plinian column 25 miles (40 km) high rose over the volcano.

SAKURA-JIMA, JAPAN
Lat. 31.58 N – Long. 130.67 E – Alt. 3,665 feet (1,117 m) – 0802-08

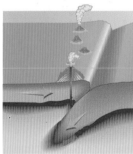

Island arc

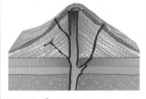

Stratovolcano

Prevalent volcanic activity. Vulcanian, lava fountains, lava flows.

General. Sakura-jima is one of the most active and dangerous volcanoes in Japan. It is located inside the Aira caldera, only 5 miles (8 km) from Kagoshima, on the island of Kyushu, one of Japan's most populous cities. The form of its cone, its position inside a large bay near a populous city and even the climate make Sakura-jima a sort of Oriental twin to Vesuvius. Even more than Vesuvius, however, Sakura-jima constitutes a constant presence in the lives of the inhabitants of its city. In fact, every year dozens of explosions rain ash over the city. The last large eruption was in 1914, when lava flows descended deep fissures on the slopes of the volcano forming an island that joined the volcano to Kyushu. Today, civil authorities and scientists use one of the world's most modern networks of volcano monitoring to keep a close eye on the volcano so as to warn the city in the case of imminent danger.

The repetitive character of the explosions has offered the scientists a chance to recognize the sequence of signals that precede an eruptive event, while continuous monitoring of wind direction matched with computer simulations enables the scientists to advise the city of where ashfall can be expected.

Thanks to these systems and to the general low level of the volcano's eruptions over the past years, the inhabitants of Kagoshima have learned to adapt their lives to the dangers presented by the nearby volcano.

Access and principal attractions. *From Kagoshima boats cross the bay to the volcano, where various pathways have been set up to view the volcano, all of which, of course, can* be used only during periods of inactivity.

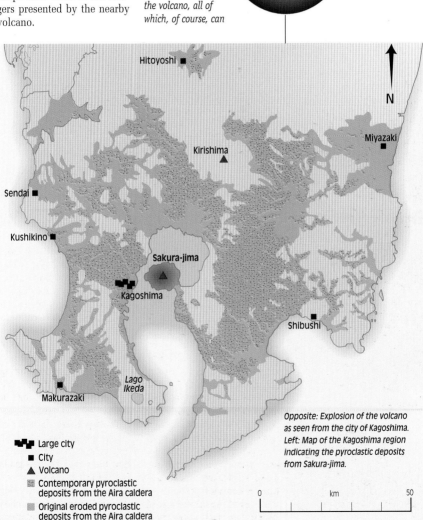

Hitoyoshi ■

N

Miyazaki ■

Kirishima ▲

Sendai ■

Kushikino ■

Sakura-jima ▲

Kagoshima

Shibushi ■

Lago Ikeda

Makurazaki ■

Legend

◆◆ Large city
■ City
▲ Volcano
▨ Contemporary pyroclastic deposits from the Aira caldera
▨ Original eroded pyroclastic deposits from the Aira caldera

Opposite: Explosion of the volcano as seen from the city of Kagoshima.
Left: Map of the Kagoshima region indicating the pyroclastic deposits from Sakura-jima.

0   km   50

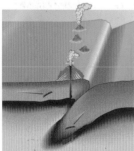

Island arc

Compound volcano

Prevalent volcanic activity. Pelean, lava domes.

General. The Mount Unzen complex is located on the Shimabara Peninsula in the northwestern part of the island of Kykushu. The compound volcano is made up of three distinct edifices grouped over an area 12½ miles (20 km) in diameter: Kinugasa, Kusenbu and Fugen-dake, the highest point of the complex. They are composed in turn of a large number of andesitic and dacitic lava domes formed on top of pyroclastic material. Three principal phases of activity have been identified in the complex, each corresponding to the formation of one of the three edifices and associated with Pelean-type eruptions (the extrusion and collapse of lava domes with the generation of *nuées ardentes*). Only a few eruptions are known to have taken place in historical time, including those of 1657 and 1792, but the latter ranks as one of the worst volcanic disasters in all of Japan, leading to the deaths of 14,524 people. It began with a series of phreatic explosions at the summit of the volcano, with dacite lava flows that extended 2¼ miles (3.5 km) down the slopes of the mountain. The eruption was not particularly intense, but three months later, likely as a result of seismic activity and also, probably, the build-up of internal pressure caused by the slow rise of magma, the easternmost lava dome, known as Mayu-yama, collapsed, causing a debris avalanche that swept 4 miles (6.5 km) down the mountain to devastate the southern quarter of the city of Shimabara, killing 9,528 people. Most of the volcanic material in the avalanche emptied into the sea, causing tsunami waves that struck the seacoast of Ariake, killing another 4,996 people. The deposits from this gigantic avalanche accumulated off the coast of Shimabara, forming numerous hummocks that have become small islands.

Following an isolated seismic crisis in 1984, in November 1989 the volcano was shaken by 400 seismic shocks over two days, from an epicenter located in the sea, 6 miles (10 km) off the coast. This seismic crisis lasted 12 months, during which the epicenter migrated progressively nearer to the volcano. In November 1990 a phase of explosive activity began from the two craters on the top of Fugen-dake, which, through a series of phreatic and phreatomagmatic explosions continued until May of the next year. On May 12 the frequency of the earthquakes increased, and on May 20 a dacitic lava dome was extruded from one of the summit craters, Jigoku-ato. Four days later a portion of this dome began to collapse, generating *nuées ardentes* (the pyroclastic flows typical of Pelean explosions). On June 3, 1991, at around 4 P.M., a larger *nuée ardent* formed and swept down a 2½-mile (4-km) channel toward the southern part of the city of Shimabara. Along the way it killed 38 people near the village of Kitakami-koba, including several journalists and three internationally known volcanologists, the American Harry Glicken and the French couple Maurice and Katia Krafft.

This type of activity continued at Unzen until February 1995; another two phases of collapse accompanied by *nuées ardentes* occurred in the months of February and May 1996.

Access and principal attractions. *The city of Shimabara, on the island of Kyushu, is the place to start. There you can organize an excursion to the numerous lava domes and the pyroclastic flows of the volcanic complex. Obviously you will need to speak to local guides and authorities and get the most up-to-date information on the state of the volcano.*

Opposite: Fugen-dake, the highest peak in the Unzen massif. Fugen-dake is composed of at least five dacitic lava domes, the easternmost of which is Mayu-yama (2,687 feet [819 m]), which produced the worst volcanic catastrophe in Japanese history.
Below: One of the highly numerous nuées ardentes ("glowing clouds") that were generated on Unzen in the years 1991-95.

Island arc

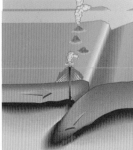

Caldera
Stratovolcano

Prevalent volcanic activity. Strombolian, Vulcanian, phreatic, lahars, fumarolic.

General. The caldera of Aso is composed of an elliptical depression (15½ miles [25 km] north-south by 11 miles [18 km] east-west) located in the central region of the island of Kyushu. One of the most active volcanoes in Japan, its activity began about 300,000 years ago with a large explosive eruption. A further three large eruptions followed, the last of which occurred between 70,000 and 80,000 years ago and gave the caldera its current shape. During this eruption pyroclastic flows covered a distance of more than 93 miles (150 km) from the crater, leaving a deposit of pumice and ash that shows up in many places over the entire island of Kyushu, in most places with a thickness of less than 6½ feet (2 m).

Inside the caldera are 15 volcanic cones arranged along an east-west axis. Only one of these cones has been active in historical time, Naka-dake, 4,941 feet (1,506 m) high, which has erupted about 170 times since A.D. 553. The peak of Naka-dake is composed of a group of craters that form an alignment 0.7 miles (1.1 km) long, one of which, Crater 1, is at an altitude of about 4,265 feet (1,300 m).

Today the southern side of Crater 1 is the site of volcanic

and fumarolic activity. Recent eruptions at Crater 1 occurred in 1979, 1984–85, 1989–90 and 1992–93.

In the period immediately preceding such eruptions, the lake inside the crater changes color, going from green to gray because of the arrival of volcanic gas. The lake then completely evaporates because of the enormous heat, which even makes part of the crater floor incandescent. Finally there are hydromagmatic and Strombolian eruptions interspersed with frequent phreatic explosions. At the end of the eruption the crater floor is again closed off and sealed by internal avalanches and the accumulation of erupted material, and a new lake forms. Since February 1993 the activity at Naka-dake has been limited to frequent explosions of mud from the crater lake, along with the opening of new fumaroles and periods of incandescence of the crater wall.

Access and principal attractions. *Aso is located at the center of a national park and can be reached by a one-hour drive from Kumamoto. There is a museum of volcanology near the entrance to the park, and a convenient road leads right up to the cone of Naka-dake. Near the cone are reinforced-cement structures where visitors can take shelter during the more intense Strombolian eruptions. Aside from these eruptions, the park also offers a beautiful volcanic landscape composed of many cones inside the large crater. You can ascend to the crater lake of Naka-dake, but doing so is not recommended because of the concentration of volcanic gas around the crater. Since 1980, 71 people have been hospitalized following inhalation of the gas, and seven of them died. During periods of greater concentrations of gas the authorities close off access to the crater.*

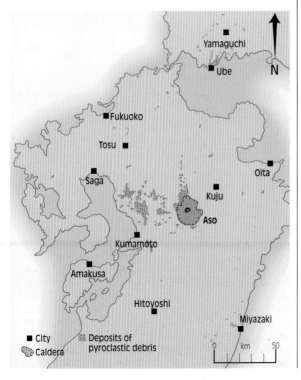

■ City
🌀 Caldera
▨ Deposits of pyroclastic debris

0   km   50

Opposite: The complex of volcanic cones inside the caldera of Aso.
Left: Map of the northern area of the island of Kykushu with the caldera of Aso and the sites where deposits of pyroclastic material from the last great eruption are visible.

FUJI, JAPAN

Lat. 35.35 N – Long. 138.73 E – Alt. 12,389 feet (3,776 m) – 0803-03 **45**

Island arc

Stratovolcano

Prevalent volcanic activity. Vulcanian, Strombolian, lava flows, lahars, fumarolic.

General. Mount Fuji, Japan's highest mountain, is located in the middle of the large island of Honshu, near its eastern coast. The volcano appears under many names: Fuji, Fuji-yama, Fuji-san, Huzi, Huzi-san. It is known throughout the world for the splendor of its symmetrical cone. At the summit is a crater 2,300 feet (700 m) in diameter and about 330 feet (100 m) deep. Between A.D. 781 and 1083, the volcano went through about 15 eruptions, both explosive and effusive (lava flows). Activity began again in 1511, but since

then there have been very few eruptions. The last of these, and also the most explosive, took place in 1707. At that time the Hoezian crater, on the southeastern flank of the volcano, erupted 0.2 cubic miles (0.8 km³) of ash and lapilli; some of this material fell on the city of Tokyo, more than 62 miles (100 km) away.

Fuji has always played a leading role in Japanese culture. Its imposing presence has given birth to many tales and legends, many still told today. One of these tells of an emperor who came into possession of an elixir of immortality; he gave orders that it be destroyed on the peak of the volcano. Since then the smoke that rises from the peak has been laden with this magical elixir, rendering anyone who inhales it immortal.

Veneration of the volcano as a sacred mountain throughout Japan began in the eighth century when a Buddhist monk made the first ascension and experienced moments of intense spirituality. Pilgrimages to the peak began during the 14th century. Today, at least 400,000 Japanese climb the volcano every year to prostrate themselves at the temple located at its peak. The mountain has also inspired much of the local artistic production, both literary and musical.

Access and principal attractions. July and August are the months for climbing Fuji because the shelters on the slopes of the volcano are open during that period. There are five main routes up the mountain: Kawaguchiko, Subashiri, Fujinomiya, Fujiyoshida and Gotemba. Each route is divided into 10 stops. The easiest route is Kawaguchiko; the route most often used for the descent is Subashiri. About nine hours are needed for the ascent; you can spend the night in one of the shelters located near the peak. Part of the route (up to the fifth stop of the Kawaguchiko, Subashiri and Fujinomiya routes) can be covered in bus or car. The panorama from the peak is superb. There is much to see

also at the foot of Fuji, including the Narusawa Ice Cave, the Fugaku Wind Cave, the enchanting Shiraito waterfalls, the panoramic hill of Koyodai and the ancient forests of Junkai and Aokigahara.

Opposite: The splendid cone of Fuji covered in snow is a recurrent image in Japanese culture.
Below: View of the 700 meter wide crater of Fuji.

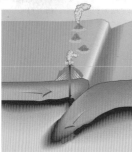

Island arc

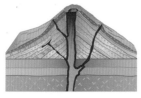

Stratovolcano

Prevalent volcanic activity. Phreatomagmatic, Plinian, Pelean, lava domes, fumaroles, phreatic.

General. Usu is a stratovolcano located on the southwestern part of the Toya caldera in the southwestern area of the island of Hokkaido. The caldera is estimated to be about 130,000 years old, but Usu was formed only about 10,000 years ago. It is a small stratovolcano with a base about 4 miles (6 km) in diameter and a maximum height of 2,398 feet (731 m). About 7,000 to 8,000 years ago the summit of the volcano broke, leaving a horseshoe-shaped depression. The volcano was then quiescent for thousands of years

until 1663, when it returned to activity with a Plinian eruption that emitted more than 0.5 cubic miles (2 km³) of pumice and ash. Since then Usu has erupted more than six times, with less intensity: in the past century in 1910, 1943–45 and 1977–78; most recently in 2000. The volcano is famous for the formation of dacitic lava domes, with associated, often phreatomagmatic, explosions and collapses of incandescent material to form pyroclastic flows. Inside the summit crater, about 1¼ miles (2 km) in diameter, are two lava domes, Ko-usu and Oo-uso, formed, respectively, during eruptions in 1663 and 1853.

Along the flanks of the volcano are another lava dome and six cryptodomes. These are masses of magma located inside the volcano and near enough its surface to change its shape. During eruptions of the last hundred years such processes have raised an area of the volcano equal to about 490 to 655 feet (150 to 200 m) with a diameter of about 0.6 miles (1 km).

In 1910 a fissure opened along the north slope at the foot of the volcano and produced a series of phreatic explosions, with the formation of 45 aligned craters. During the last two eruptions, a series of phreatomagmatic explosions led to the development of new fields of fumarolic activity. In 1977–78 the phreatomagmatic activity was preceded by an explosive magmatic phase with the formation of pumice.

Current activity at the volcano has been limited to abundant fumarolic emissions, seismic swarms and notable terrain deformation, all believed to be connected to the subterranean movement of masses of high-viscosity magma.

Access and principal attractions. Located halfway between Sapporo and Hakodate, in the Shikotsu-toya national park, Usu is easily reachable from nearby Lake Toya along a road that ends just a few miles from the summit. Travel agencies in the cities of Sapporo and Hakodate offer guided tours during which you can visit the alignment of phreatic craters formed on the north slope in 1910 as well as the fumarolic fields in the summit area.

Opposite: Dacitic lava domes in the summit area of the volcano, covered by pyroclastic material from the eruption of 1977–78.
Left: Map showing the location of Usu in the southwestern part of the island of Hokkaido near Lake Toya.

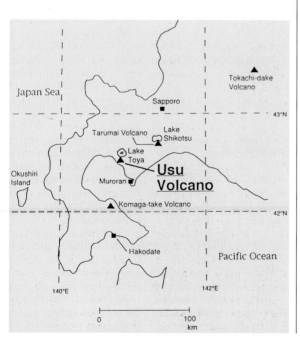

Island arc

Stratovolcano

Prevalent volcanic activity. Plinian, pyroclastic flows, Strombolian, lava flows.

General. Karymsky is the most active volcano of eastern Kamchatka. Its symmetrical cone is part of a volcanic complex located inside the Karymsky caldera, about 22 by 40 miles (35 by 65 km) in size, formed about 7,700 years ago following a large explosive eruption. The activity of Karymsky can be divided into three large periods, 6,100–5,100 years ago, 4,300–2,800 years ago and from 500 years ago to today. The first period was one of very intense activity, with the formation of pyroclastic deposits of various composition (from basaltic andesite to dacite) that can be found up to 6 miles (10 km) from today's crater. Later activity was more moderate, with the emission of lava and pyroclastic products of andesitic composition.

As for the most recent period of activity, still in course (with an eruption in 2000), it has involved the emission of lava flows of andesitic composition and persistent Strombolian activity.

The current state of the volcano is very similar to that of Stromboli; beginning on January 2, 1996, there were explosions of volcanic gas and ash that repeated with regularity, every 20–40 minutes, with high points of 100 eruptive events daily. The material erupted during these explosions was thrown up to heights of 655 to 1,310 feet (200 to 400 m) above the caldera. Larger explosions occasionally produced columns of gas and ash that reached 2 to 2½ miles (3 to 4 km) in height. Because

of this extremely intense eruptive activity, practically the entire cone of the volcano is covered by lava flows from the past 200 years. The geometry and dimension of the summit crater change continually from eruption to eruption.

Access and principal attractions. *Karymsky is located in a remote region without any overland access. You can reach it only by helicopter, leaving from Petropavlovsk, the seaport that is the capital city of Kamchatka. The flight takes about two hours. Flying over the volcano you see its wilderness setting, an unspoiled region studded with numerous volcanic cones and craters. The helicopter can land at the foot of the volcano, from where you can witness Strombolian activity.*

Opposite: Snow on the summit of the volcano, seen from the north. Behind it is the caldera lake of Odnoboky, also part of the Karymsky complex.
Above: The cone of Karymsky in summertime.
Left: Aerial view of the interior of the summit crater.

208

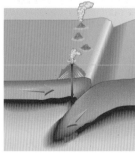

Island arc

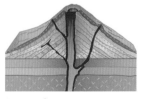

Stratovolcano

Prevalent volcanic activity. Vulcanian, Pelean, lava flows, lava domes, fumarolic.

General. Bezymianny, located in central Kamchatka, is part of a group of volcanoes known as the Kljucevskaia alignment. This volcano is best known for the eruption of 1955–56, which ended about 1,000 years of quiescence and was one of the largest Pelean events of historical time. Because of the volcano's remote location there were no fatalities, and in fact the eruption drew only modest attention from the press at the time.

The eruption began on October 22, 1955, its coming announced by a series of earthquakes that lasted 23 days. At the end of November a lava dome began forming on the bottom of the crater, accompanied by notable swelling of the volcano's southern slope.

The volcano produced numerous Vulcanian explosions, with jets of cinders and gas up to heights of 4⅓ to 5 miles (7 to 8 km), until March 30, 1956. Suddenly, on that day, there was a large lateral explosion of the lava mass of the dome, accompanied by an avalanche along the eastern slope, followed by a Plinian phase. An eruption column of gas and ash rose roughly 25 miles (40 km) into the atmosphere. The eruption caused the formation of a horseshoe-shaped depression about 1 mile (1.5 km) wide; when the activity ended the volcano itself was about 985 feet (300 m) lower. The pyroclastic flow produced by the explosion of March 30 moved with great violence, devastating an area of 193 square miles (500 km^2), fortunately uninhabited. Studies made during the 1980 eruption of Mount St. Helens (U.S.A.) reveal extraordinary similarities with that of the March 30, 1956, eruption of Bezymianny, bringing to light an eruptive dynamic of great danger that has been recognized to be recurrent in volcanoes of convergent margins that erupt andesitic and dacitic lava.

After the eruption, Bezymianny entered a phrase of persistent activity during which a new lava dome was formed inside the depression. The growth of the dome was accompanied by numerous explosions. An even larger explosion took place in 1985, when the pyroclastic flow descended more than 8 miles (13 km) of distance from the crater. Since 1977 lava extruded from the summit of the dome has formed lava flows along the eastern flank.

Access and principal attractions. Like Karymsky, Bezymianny can be reached only by helicopter from Petropavlovsk (four to five hours). The overflight offers views of the caldera and the lava dome inside it.

Kamchatka Peninsula

Bezymianny

Karymsky

Opala   PETROPAVLOVSK

0   100   200 Km

Opposite: The wall of the depression formed following the 1956 eruption. Near the center is the lava dome that has been forming in the years since then. The dark area visible to the left of the dome is a lava flow, one of the many that have been discharged periodically from the dome since 1977.
Left: Map of the Kamchatka Peninsula with Bezymianny and Karmysky. The red triangles indicate active volcanoes, of which there is an enormous concentration in the area.

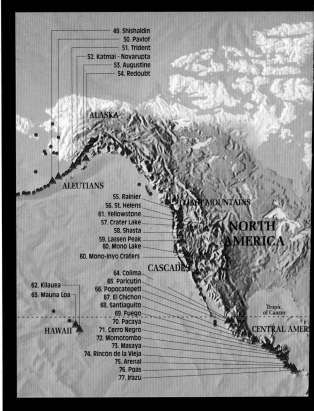

49. Shishaldin
50. Pavlof
51. Trident
52. Katmai - Novarupta
53. Augustine
54. Redoubt

ALASKA

ALEUTIANS

55. Rainier
56. St. Helens
61. Yellowstone
57. Crater Lake
58. Shasta
59. Lassen Peak
60. Mono Lake
60. Mono-Inyo Craters

COAST MOUNTAINS

NORTH
AMERICA

CASCADES

64. Colima
65. Paricutin
66. Popocatepetl
67. El Chichon
68. Santiaguito
69. Fuego
70. Pacaya
71. Cerro Negro
72. Momotombo
73. Masaya
74. Rincón de la Vieja
75. Arenal
76. Poas
77. Irazú

62. Kilauea
63. Mauna Loa

HAWAII

Tropic
of Cancer

CENTRAL AMER

Almost the entire length of the Pacific rim of both American continents, from the arc of the Aleutians in Alaska to Tierra del Fuego at the
southernmost tip of South America, is the scene of volcanic activity.
Other active volcanoes are grouped in the island arcs of the West Indies
or are located inside plates (the Hawaiian Islands, the Galápagos and
various volcanic centers scattered across the western United States).

Volcanoes of the American continental margin

From north to south the most active volcanic centers are concentrated along the arc of the Aleutians (44 volcanoes) and on the Alaskan peninsula (53 volcanoes); along the Coast and Cascades mountain ranges, which extend from Canada to California (72 volcanoes); from the vol-canic area of Mexico, which includes the active centers of Baja California (35 volcanoes), to the interior of Central America from Guatemala to Panama (74 volcanoes); and then in South America the Andes arc, which is found in the three regions Colombia-Ecuador (34 volcanoes), Peru-Bolivia-Chile (82 volcanoes) and, farther south, Chile-Argentina

(67 volcanoes). During the last century there were large explosive eruptions at Mount Pelée (Martinique, 1902), Katmai-Novarupta (Alaska, 1912), St. Helens (Cascades, 1980), El Chichón (Mexico, 1982), Nevado del Ruiz (Colombia, 1985) and Cerro Hudson (Chile, 1991). The dominant volcanic activity in all these areas is explosive (Strombolian, Vulcanian, Plinian), and the most frequently encountered types of volcanoes are stratovolcanoes and calderas, sometimes of exceptional size. In the course of the eruptions the snow covering of peaks or the glaciers that crown volcanic cones at medium or high altitudes can be subject to sudden melting, leading to flows of mud and detritus (lahars).

Volcanoes inside plates
The western regions of the United States, inside the continental plate, are home to 20 active volcanic centers; these are located in the states of Idaho, Wyoming, Nevada, Utah, Colorado, Arizona and New Mexico. Also included in this group is the hot spot of Yellowstone, site of the famous geysers, and Arizona's Sunset Crater, the only one to have documented eruptive activity in historical time (around the 13th century). Inside the Pacific oceanic plate, roughly 3,100 miles (5,000 km) off the coast of Mexico, is the hot spot of Hawaii, site of some of the world's most active and famous volcanoes. Hawaii has a total of 13 active volcanoes, some of them submerged. For more than 15 years Kilauea has been in perennial eruption and with Mauna Loa has had numerous eruptions in recent centuries. These eruptions produce basaltic alkaline-sodic magma in lava fountains and lava flows that cover dozens of miles before emptying into the ocean. Another hot spot is off the coast of Ecuador near the Galápagos Islands, inside the Nazca oceanic plate. There are 15 active volcanoes there, nine of which erupted several times during the last century, emitting basaltic lava flows.

The arc of the Lesser Antilles (West Indies)
The arc of the Lesser Antilles (16 volcanoes), located on the eastern edge of the Caribbean area, is produced by the process of subduction of the Atlantic plate under the Caribbean plate. Several of the major eruptions of the last century took place in that region, without doubt the most famous being the 1902 eruption of Mount Pelée, which killed 28,000 people in the city of St. Pierre.

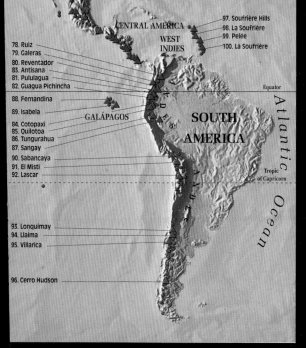

78. Ruiz
79. Galeras
80. Reventador
83. Antisana
81. Pululagua
82. Guagua Pichincha
88. Fernandina
89. Isabela
84. Cotopaxi
85. Quilotoa
86. Tungurahua
87. Sangay
90. Sabancaya
91. El Misti
92. Lascar
93. Lonquimay
94. Llaima
95. Villarica
96. Cerro Hudson

97. Soufrière Hills
98. La Soufrière
99. Pelée
100. La Soufrière

CENTRAL AMERICA
WEST INDIES
GALÁPAGOS
SOUTH AMERICA
Equator
Tropic of Capricorn
Atlantic Ocean

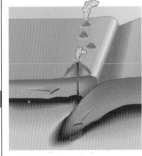

Island arc

Stratovolcano

Prevalent volcanic activity. Strombolian, lava flows.

General. Shishaldin, the highest volcano in the Aleutians, has a superb symmetrical cone that rises over Unimak Island, 37 miles (60 km) east-northeast of Westdahl and 56 miles (90 km) west-southwest of Cold Bay (Fort Randall). The cone, with an overall volume of 72 cubic miles (300 km³), is truncated at the summit by a small crater from which a plume of smoke always rises. Because of its remote location, the volcano is not well known. It is also true that observations of its activity have often been blocked by

212

cloud cover. Thirty-seven eruptions have been recorded since 1775, with an average of six years' interval between them. The explosive activity is prevalently Strombolian, with eruptions that usually last several hours. The events that occurred between 1986 and 1987, however, went on for several months. During the 1800s and into the 1900s until 1932, eight lava flows were given off, most of them from vents that opened in the summit area of the cone.

Major eruptions occurred in 1830 and 1932, both involving explosive activity with the emission of lava. Between 6:30 and 8:00 on the evening of December 23, 1995, pilots flying over the area observed the formation of a column of ash and smoke 6½ miles (10.5 km) high, elongated to the northeast by the wind. Later, at 1:30 a.m. on December 24, a light rain of ash fell in the area of Cold Bay, about 55 miles (90 km) to the northeast of the volcano. Lahars were recorded in 1922 and 1932. There was further eruptive activity in May of 2000.

Opposite: Aerial view of the cone of Shishaldin. Gas and water vapor are always being given off.
Below: Map showing the main volcanoes in the Aleutians. Volcanoes exist here because of the subduction of the Pacific plate beneath that of the Bering Sea.

Access and principal attractions.

Shishaldin is located on the eastern end of the Aleutians, not an easy place to get to; even Anchorage is a long way off. That city, however, is the place to organize a trip to the volcano, whether by helicopter with a guide or in one of the fly-bys offered to tourists. The splendid symmetrical cone, the a'a lava flows and the untouched natural beauty of the Aleutians are the main reasons for making the trip.

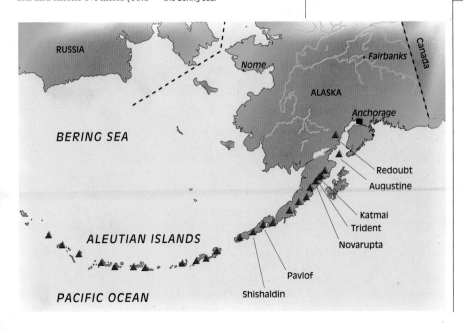

RUSSIA

Nome

Fairbanks

Canada

ALASKA

Anchorage

BERING SEA

Redoubt

Augustine

Katmai

Trident

Novarupta

ALEUTIAN ISLANDS

Pavlof

Shishaldin

PACIFIC OCEAN

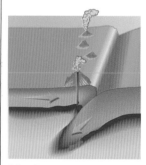

Island arc

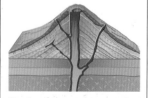

Stratovolcano

Prevalent volcanic activity. Vulcanian, Strombolian, lava flows.

General. Pavlof, the most active volcano in the Aleutian arc, is a conical stratovolcano with a diameter of 4⅓ miles (7 km), most of it covered in snow. There are eruptive vents to the north and east of the summit. The volcano is located 620 miles (1,000 km) to the southwest of Anchorage. Since 1790, when recording of the volcano's activity began, there have been 40 eruptions. The volcano has periods of rest that last from three to eight years. Some of its eruptions are brief (one or two days); others have gone on sporadically for many years. The most important event lasted about five years, from 1906 to 1911, with Strombolian explosions from the summit crater and emissions of lava from a fissure vent that opened along the north flank of the cone. Other eruptions took place in 1922–23, 1929–31, 1936–48 (12 years), 1960–63, 1975–77 and 1986–88. These usually begin with powerful Vulcanian explosions and continue with Strombolian activity and the emission of lava flows. Pavlof's largest historical eruption took place on December 6–7, 1911, near the end of an eruptive cycle that had been going on for five years. The opening of the vent on the north flank of the cone produced a lava flow and set off strong explosions in the central crater, with the ejection of large blocks of rock. During the events of 1996 there were lava fountains 655 feet (200 m) high, described as reddish plumes by the inhabitants of Cold Bay, about 31 miles (50 km) away. During the first days of October the lava fountains were joined by a lava flow down the northwest flank of the volcano. Pavlof reached the highest level of activity in mid-October with plumes of ash that rose 3¾ to 4⅓ miles (6 to 7 km) and spread out for more than 31 miles (50 km) to the southwest. Satellite images taken during November revealed that the plume of ash had reached a distance of 155 miles (250 km) from its origin on the volcano.

Access and principal attractions. To reach this distant area of Alaska you should once again begin in Anchorage, where a variety of trips can be organized with any one of the many local travel agencies. There are helicopters to take you to the volcano, guided tours and fly-bys in special aircraft. The major attractions, aside from the wonderful natural beauty of the Alaska peninsula, are Strombolian activity and lava flows.

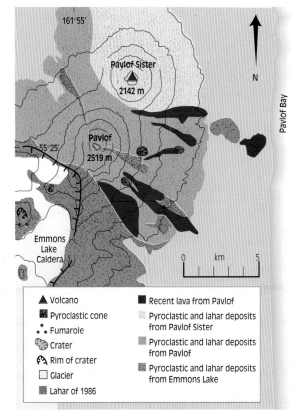

161°55'

Pavlof Sister

2142 m

N

Pavlof

55°25'

2519 m

Pavlof Bay

Emmons
Lake
Caldera

0   km   5

▲ Volcano

Pyroclastic cone

•⋮• Fumarole

Crater

Rim of crater

☐ Glacier

Lahar of 1986

■ Recent lava from Pavlof

Pyroclastic and lahar deposits
from Pavlof Sister

Pyroclastic and lahar deposits
from Pavlof

Pyroclastic and lahar deposits
from Emmons Lake

Opposite: The cone of Pavlof, at the center of the photograph, as seen from the south. Behind it is Pavlof Sister; in the foreground is Mount Hague.
Left: Map of the volcanic chain of Pavlof and Pavlof Sister, its twin volcano to the northeast, with the Emmons Lake caldera.

TRIDENT, ALASKA, U.S.A.

Lat. 58.23 N – Long. 155.08 W – Alt. 6,116 feet (1,864 m) – 1102-16

Continental margin

Stratovolcano

Prevalent volcanic activity. Strombolian, lava flows, fumaroles.

General. The area around the broad glacial Valley of Ten Thousand Smokes, the site of pyroclastic flows in 1912 from the eruption of Novarupta, is surrounded by a high concentration of young volcanic peaks, indications of the high production of magma in the mantle in that area. This intense magmatic activity is a result of the subduction of the Pacific oceanic plate under the continental shelf of Canada.

Trident, located 2 to 3 miles (3 to 5 km) to the southeast of the vent of Novarupta, stands between the active volcanoes of Mounts Mageik to the southwest and Katmai to the northeast. When it was named, Trident was composed of an elongated rise formed from eruptive materials from three distinct cones, which existed in that area before 1953. The volcanic complex also included two lava domes on its southern flank, both of them partially destroyed by glacial erosion. During the 1953 event a fourth cone, which today has a height of 1,310 to 2,625 feet (400 to 800 m), was added to the southwestern flank.

The 1953 event began with the emission of a column of ash and vapor to a height of about 5½ miles (9 km); it was emitted from a vent that opened at 3,600 feet (1,100 m) of altitude in the area of a fumarole field. The eruption continued, with phases of activity alternating with periods of inactivity, until 1960, leading to the creation of a lava field composed of many overlapping flows that extended 2¾ miles (4.5 km) from the vent. The cinder cone formed at the center of this emission continued to grow through the addition of material thrown up by Strombolian explosions and overflows of lava until 1963. Over those 10 years of activity (1953–63) a volume of material had been emitted roughly equal to a little less than 0.1 cubic miles (0.5 km³).

The summit crater of the cone, open toward the southwest, today measures 820 feet (250 m) in diameter and is the site of important fumarolic activity. In 1979 the fumaroles in this area all had the same temperature as that of boiling water at the same altitude (207°F [97°C]); some of these were rich in sulfur dioxide, creating sulfur deposits. There were also dozens of freshwater springs, which had come into being along the edge of the lava field from 1953–60; these had temperatures that varied from 54°F to 122°F (12°C to 50°C).

Today the Trident edifice has a volume of 2.4 to 3.6 cubic miles (10 to 15 km³), but it seems likely that at least half of the structure has been destroyed by glacial erosion, which is very intense in this area.

Access and principal attractions.
Trident is located in the Katmai National Park, an uninhabited wilderness area. It can be reached from Anchorage stopping first at King Salmon, about 280 miles (450 km) to the southwest, where you can take a small seaplane to Brooks Lodge. From there you face a 12½-mile (20-km) hike up the Valley of Ten Thousand Smokes. Of course, the volcano can also be reached with a day's hike from Katami Bay. Possible dangers include encounters with bears; among the major difficulties is the crossing of streams.

217

Opposite: One of the cones that form the Trident complex. In the foreground is a field of pumice and ignimbrite created by the large 1912 eruption of Novarupta.
Below: Map of the volcanic peaks around the Valley of Ten Thousand Smokes with the eruptive vent of Novarupta at the center.

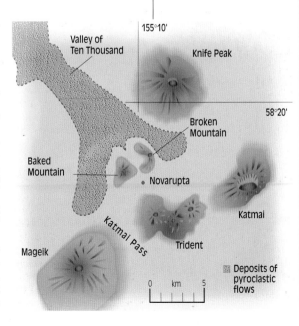

155°10'

Valley of Ten Thousand

Knife Peak

58°20'

Broken Mountain

Baked Mountain

• Novarupta

Katmai

Katmai Pass

Trident

Mageik

0   km   5

Deposits of pyroclastic flows

Continental margin

Prevalent volcanic activity. Plinian, pyroclastic flows, lava domes.

General. The volcanoes of Novarupta and Katmai are located on the Alaska peninsula near the continental margin in an almost inaccessible region facing Kodiak Island. Their eruption of June 6, 1912, which ranks among the most violent of the past hundred years. The tremendous explosion that began the eruptive activity was heard within a range of 745 miles (1,200 km) and threw such an enormous quantity of ash into the atmosphere that all of the area of Kodiak was shrouded in darkness for more than 60 hours. The ash cloud moved: by June 9 ash was falling on Vancouver and Seattle; by June 10 it had passed Virginia, and seven days later it had reached Algeria. Because of the remote location, the direct effects of this enormous volcanic event remained virtually unknown outside the area. The first scientific expedition, organized by the National Geographic Society a month and a half after the event, found that the peak of Katmai was completely gone; in its place there was a caldera 2 by 2½ miles (3 by 4 km) in size and between 1,970 and 3,610 feet (600 and 1,100 m) deep. A later expedition, in 1916, identified the valley where most of the material from the eruption had accumulated. Because of the enormous number of fumaroles releasing plumes of steam from the still-hot terrain, they named the valley "Ten Thousand Smokes." Later studies revealed that Katmai had not been involved in the eruption; the volcanic center responsible for the eruption, called Novarupta, was about 5 miles (8 km) west of the Katmai caldera. During the course of the eruption the magma beneath Katmai had been "sucked" away from it by Novarupta through a large fissure that joined the two

places reached depths of more than 165 feet (50 m). The scientific study of this eruption proved of great importance to the field of volcanology; for the first time it was possible to prove that large explosive eruptions can produce pyroclastic flows, avalanches of incandescent pumice and ash mixed with volcanic gas. These often flow into valleys and are capable of traveling many miles from the eruptive vent

Access and principal attractions. Getting to Katmai National Park begins with a flight from Anchorage to the city of King Salmon, roughly 280 miles (450 km) away. From there you reach Brooks Lodge in a small seaplane. You can then drive to the Valley of Ten Thousand Smokes, with its enormous ignimbrite deposits cut by streams. The rest of the visit, including an examination of the valley and visits to Novarupta (and its lava dome) and the Katmai caldera, can be made only on foot. The national park, with a surface area of 4,218 square miles (10,925 km²), was created in 1918 to preserve and protect the local flora and fauna, in particular the brown bear.

volcanic systems. Novarupta had initially erupted rhyolitic magma from its own magma chamber, and then dacitic and andesitic magma drawn from the magma chamber of Katmai. The eruption had covered a large area with the particular kind of pyroclastic deposit known as ignimbrite; the amount involved was colossal, about 6 cubic miles (25 km³) of material, which in some

(in this case more than 9½ miles [15 km]). After the paroxysmal phase of the eruption, the activity inside the crater of Novarupta continued with the formation of a lava dome of rhyolitic composition 295 feet (90 m) high and 1,180 feet (360 m) wide. Since 1912 there have been no more reports of eruptions from the Katmai-Novarupta system.

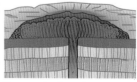

Novarupta: lava dome

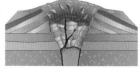

Katmai: caldera

220

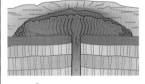

Continental margin

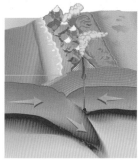

Lava dome

Prevalent volcanic activity. Vulcanian, Pelean, pyroclastic flows, lava flows, lava domes, lahars, fumaroles.

General. Augustine, one of the youngest and most active volcanoes in the Alaska-Aleutian arc, forms an island located at the opening to Cook Inlet, about 175 miles (280 km) to the southwest of Anchorage. In fact it was James Cook who named the island, having first sighted it on the feast day of St. Augustine in 1778. It has a regular conical shape, although the presence of a lava dome at the summit gives it a round appearance. Extensive accumulations of detritus have given the volcano gentle slopes, but the central area of the island is composed of short lava flows on a somewhat steep slope. Aside from lava, Augustine erupts bread-crust bombs, pumice and ash of both andesitic and dacitic composition. The magma originates at a depth of about 62 miles (100 km) where there is a subduction plane resulting from the Pacific plate moving under the Canadian Shield.

Seven eruptions have occurred here over the course of the last two centuries; in 1812, 1883, 1935, 1963–64, 1971, 1976 and 1986, with intervals of rest varying in length from five to seventy years. These eruptions have tended to be explosive, spreading volcanic ash throughout the entire region of Cook Inlet. The explosions that took place during the last days of January 1976 produced columns of ash and gas up to 6 miles (10 km) high that caused a light rain of ash on the city of Anchorage.

In many cases pyroclastic flows have been generated that descended the slopes of the volcano in all directions. The end of an eruption is usually marked by the slow extrusion of a lava dome at the summit of the cone; such cones are eventually destroyed during the next Pelean explosion. The extrusion of the dome increases the steepness of the slopes, thus favoring the formation of avalanches. In fact, Augustine is surrounded by vast detritus fields studded with hummocks produced by such avalanches. In the past, the sudden entry into the sea of an avalanche has produced tsunami waves, which on several occasions struck the coasts of Cook Inlet. Small tsunami waves have been produced by pyroclastic flows pouring into the sea. Explosive eruptions that take place during the winter melt the snow, producing lahars.

Access and principal attractions. *Because of its high level of volcanic activity and its remote location, the island of Augustine is difficult to visit. Anyone planning a visit should first get in touch with the Alaska Volcano Observatory in Fairbanks to find out the volcano's current situation. Flights over the site can be made from Anchorage, and they offer the best chance to see the volcano in its wonderful natural setting.*

221

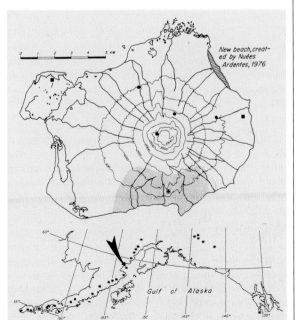

New beach, created by Nuées Ardentes, 1976

Gulf of Alaska

Opposite: Augustine seen from the north, surmounted by the plume of a large fumarole. The foreground is studded with many hummocks, the characteristic formations following volcanic avalanches.
Left: Map of the island, which is located at the opening to Cook Inlet; altitudes are given in feet.
Above: Winter view of Augustine.

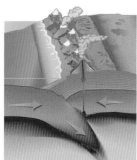

Continental margin

Stratovolcano

Prevalent volcanic activity. Vulcanian, Pelean, pyroclastic flows, lava domes, lahars.

General. One of the most active volcanoes in Alaska, Redoubt is located 105 miles (170 km) to the southwest of Anchorage and about 12½ miles (20 km) from the coast of Cook Inlet in Lake Clark National Park. The volcano began activity about 900,000 years ago and has a diameter of about 9⅓ miles (15 km); its cone, the result of lava flows, is partially covered by ice- and snowfields. The volcano is located between two rivers, the Drift River to the north and the Crescent River to the south, and most of the mudflows produced following eruptions empty into these rivers. At the summit is a horseshoe-shaped depression open to the north, probably the result of an ancient avalanche. Studies of deposits indicate that the volcano has erupted at least 20 times over the past 10,000 years. Before the event in 1989–90, the volcano's history included an eruption in 1902 and several explosions in 1966 and 1967.

From December 1989 to April 1990, Redoubt was the scene of a series of explosive episodes, with the formation of clouds of gas and ash that rose up to 7½ miles (12 km) in the air. The explosions were followed by a phase of slow emission of lava, with the formation of domes around the summit crater. Very few people live in the immediate vicinity of the volcano, but the lahars that resulted from the snow and ice melted by the explosion on December 14, 1989, caused a total of $160 million in damages to petroleum installations near the fork of the Drift River. On December 15 a cloud of gas and ash rose 7½ miles (12 km) in the air and struck a jet plane flying over the area, causing temporary engine failure and leading to an emergency landing at Anchorage airport.

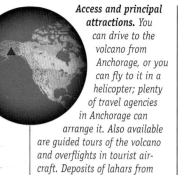

Access and principal attractions. *You can drive to the volcano from Anchorage, or you can fly to it in a helicopter; plenty of travel agencies in Anchorage can arrange it. Also available are guided tours of the volcano and overflights in tourist aircraft. Deposits of lahars from December 1989 are visible in the Drift River Valley in the splendid natural setting of the Alaska peninsula.*

Opposite: The volcano while giving off a dense cloud of gas and steam from its summit crater. In the foreground are glacial valleys that spread outward from the peak. Left: Redoubt, topped by a high plume of gas and smoke, seen from the west following the eruptions of 1989–90.

Continental margin

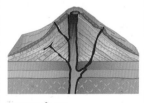

Stratovolcano

Prevalent volcanic activity. Plinian, lava domes, lava flows, mudflows.

General. With its 14,410 feet (4,392 m) of altitude, Mount Rainier is the highest volcano in the entire Cascades chain; on clear days it is visible from more than 93 miles (150 km) away. It is located a little under 50 miles (80 km) from Mount St. Helens, which has recently surpassed it in international fame, although Rainier, with an age of 730,000 years, is the much older mountain. Built up from the accumulation of lava released from the center of the edifice, the volcano's flanks are today covered by 27 glaciers spreading out radially around the peak. In some places this layer of snow and ice is more than 55 feet (17 m) deep. If the volcano were to erupt, this enormous mass would melt and become a giant and terrible lahar, certain to cause enormous damage to the valleys running along the flanks of the volcano. About 500 years ago an eruption melted a large quantity of snow, setting off a mud landslide today known as the Electron Mudflow. This spilled into nearby Puyallup Valley, uprooting and destroying trees and plants and filling the valley floor with a mass of detritus that in some places is more than 20 feet (6 m) thick. Lahars seem to be a distinctive trait of Rainier's activity; traces of at least 12 such events have been found dating back 5,600 years, to the date of the most violent mudflow, known as the Osceola Mudflow. It swept downhill at a velocity of 100 miles (160 km) per hour and covered a distance of more than 62 miles (100 km).

The dramatic cone of the volcano, known in the local Native American language as Tacoma (the name applied to the nearby city), was shaped by alternating layers from flows of andesitic lava and tuff. The most recent explosive eruption dates to 2,200 years ago. It was Plinian; most of the pumice and ash emitted was carried to the east by the wind or fell to form a layer of pumice at the foot of the volcano that is more than 1 foot (30 cm) thick.

Several small avalanches are known to have occurred in the 19th century: in 1843, 1854, 1858, 1870, 1879 and 1882. Such events must have had an impact on the population in the area, and events like these show up in stories told by local Native American tribes.

Volcanologists believe that Mount Rainier constitutes a potential danger since it seems probable that its current state of quiescence will end with a large-scale explosive eruption.

Opposite: Mount Rainier seen from the east in summer.
Below: The cone of Mount Rainier in the light of sunset.

Access and principal attractions. *Mount Rainier is located in Washington State about 40 miles (65 km) to the southeast of Seattle. The volcano is inside its own national park, founded in 1899, between Highways 5 and 90. The best—and most spectacular—way to reach the park is from the west, along Route 706 from Nisqually and Paradise. This is one of the most panoramic and beautiful roadways in the United States. The many attractions inside the park include the Trail of the Shadows, Wonderland Trail, Christine Falls and Ricksecker Point Road, with its landscapes of the Nisqually glacier.*

226

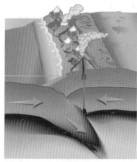

Continental margin

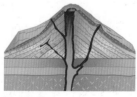

Stratovolcano

Prevalent volcanic activity. Plinian, pyroclastic flows, lava flows, lava domes, lahars. **General.** The formation of Mount St. Helens began in the Pleistocene, and much of its massif was produced by basaltic and andesitic lava flows. The summit area, or what remains of it, is 2,200 years old. As many as 60 layers of tephra have been distinguished, documenting the volcano's frequent explosive activity in the past, probably the most active volcano in the Cascade Range during the Holocene. Before the major eruption in 1980, St. Helens was composed of a regular cone, with a base diameter of about 4⅓ miles (7 km), that reached an overall height of almost 9,840 feet (3,000 m). Five large glaciers covered its peak. The surrounding area, with its immense stands of ancient conifers and the banks of Spirit Lake at the foot of the volcano, was a popular vacation spot for thousands of Americans. Numerous eruptions had been witnessed by Native Americans (the volcano appeared under various names in their stories and legends), and early settlers had seen eruptions in the 19th century.

On March 20, 1980, a series of medium-intensity earthquakes signaled the imminent return to activity. Small eruptions followed one another for about two months, during which period the northern flank of the mountain progressively lifted about 490 feet (150 m) under the force of the magma rising inside the volcano. At 8:32 A.M. on May 18, a high-intensity surface earthquake shook the entire edifice, causing the collapse in three blocks of the northern slope and at the same time the explosion of the magma accumulated inside the mountain. The avalanche and associated explosion decapitated the cone, creating a crater open to the north, 1,970 feet (600 m) deep and 1¼ miles (2 km) in diameter. The cloud of gas and debris blasted northward by the explosion—dubbed the "Stone Wind" in the media—traveled at supersonic speed at an estimated temperature of 482°F (250°C), devastating almost 232 square miles (600 km²) of forestland. A vertical column of gas and ash formed over the volcano, in a few hours reaching a height of about 10 miles (16 km). Around noon a series of pyroclastic flows descended along the valleys of the volcano, and in the afternoon a Plinian column rose to a height of about 12½ miles (20 km). Around 5 P.M. the paroxysmal phase of the eruption ended, leaving a destroyed landscape and a gutted mountain. Following the eruption, a series of lahars added further damage to the surrounding territory, causing devastation in many of the river valleys that spread out from the volcano.

After the May 18 eruption, St. Helens erupted in an explosive manner five more times that year. None of these eruptions was equal in intensity to the first, but each produced columns of gas and ash that reached heights between 5 and 10 miles (8 and 16 km). There were also pyroclastic flows down the northern slope of the volcano, some of which traveled at speeds of 62 miles (100 km) per hour. Deposits from these pyroclastic flows show up even 5 miles (8 km) to the north of the volcano.

A few days after explosive events on June 12 and August 7, lava of the same dacitic composition as that of the material ejected during the explosions was discharged

Opposite: Summit area of Mount St. Helens following the eruption of May 1980, with the large crater blown open to the north. Smoke rises from a lava dome that formed inside the crater during the months after the eruption.
Below: The cone of the mountain before the 1980 eruption, as seen from the northeast. In the foreground is a view of Spirit Lake.

Above: The eruption of July 22, 1980. The cloud of gas and ash expands upward, forming a region of convective currents from which solid material falls to the ground. At the same time, part of the column collapses under its own weight, forming a pyroclastic flow that discharges from the crater down the opening in the north side and flows down valleys near the volcano.

into the crater of the volcano. Too viscous to flow, this lava formed domes that were then destroyed by later explosions. The lava dome that formed following explosive events in October 1980 stood up to later explosions and eventually formed the nucleus of the current dome. The events in October were followed by 11 more eruptions, most of them effusive and all of them contributing to the growth of this new lava dome. Many of these eruptions were preceded by small explosions. When such eruptions have occurred in winter months, the heat released by the eruption has melted part of the

snow cover, setting off lahars and snow avalanches.

During the explosive event on March 19, 1982, pumice and rock from the dome were thrown against the southern wall, causing an avalanche of snow and rock that crossed the length of the crater to spill out the hole on the north side opened by the major 1980 eruption. Much of this reached the course of the North Fork Toutle River, causing further destruction. Vigorous emissions of gas and ash occasionally occur at the top of the lava dome, producing volcanic plumes $3\frac{3}{4}$ to $4\frac{1}{3}$ miles (6 to 7 km) high.

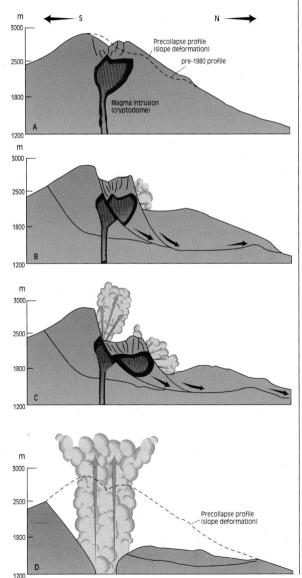

m
3000
2500
1800
1200
A

S ← → N

Precollapse profile
(slope deformation)

pre-1980 profile

Magma intrusion
(cryptodome)

m
3000
2500
1800
1200
B

m
3000
2500
1800
1200
C

m
3000
2500
1800
1200
D

Precollapse profile
(slope deformation)

Access and principal attractions. *Located in Washington State, the protected area of Mount St. Helens National Volcanic Monument is well designed from the tourist's point of view. A visitors' center is located on Wash-504 near Castle Rock. You can participate in guided tours led by rangers along various nature trails in the northern sector. There are also many local agencies that offer spectacular helicopter flights, including overflights of the interior of the crater. Other attractions of a visit are the volcano torn open by the 1980 eruption, with its many volcanic deposits; Spirit Lake, partially covered by enormous trees knocked down by the eruption; and forests completely leveled by the force of the initial blast.*

Left, from top to bottom: The eruption of May 18, 1980. In the first phase (A), magma inside the volcanic edifice forms a growing sack, pushing outward and causing the northern flank of the mountain to swell. In the second and third phases (B, C), following a surface earthquake, the northern flank of the mountain breaks, exposing the magma sack (cryptodome), causing its immediate explosion to the north. The debris-laden cloud produced is projected at supersonic speeds over distances up to 10 miles (15 km) from the volcano. In the final phase (D), a vertical column of gas, pumice and ash forms as the eruption enters its Plinian phase.

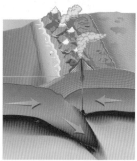

Continental margin

Caldera

Prevalent volcanic activity. Plinian, pyroclastic flows, caldera collapse.

General. Crater Lake, which gives its name to a national park in Oregon, is composed of a large caldera partially filled by a lake. Considered one of the most spectacular natural beauties of the United States, the lake is a singularly intense shade of blue, for which reason it has been dubbed the "shining jewel of the Cascades."

The caldera, circular in form and 6 miles (10 km) wide, was created about 6,800 years ago following the enormous explosion of a stratovolcano, known as Mount Mazama, which stood about 3,280 feet (1,000 m) higher than today's caldera. This eruption is considered one of the major eruptions of the last 10,000 years of our planet's geological history. The eruption can be divided into two stages. In the first, magma was erupted from a central crater, and a gigantic Plinian column rose and spread, dispersing its ash over an enormous area. The modern-day American states of Oregon, Washington, Idaho and Montana as well as a large part of the Canadian province of Alberta were covered. The column then collapsed under its own weight, producing pyroclastic flows whose deposits, known today as the Wineglass Welded Tuff, formed spectacular outcroppings on the flanks of the volcano. The second stage involved many pyroclastic flows from a series of fissures that opened along the volcanic cone and led to the cone's collapse. By the end of the eruption an estimated 12 cubic miles (50 km³) of magma had been emitted, and the enormous caldera had been formed. It then came to be occupied by Crater Lake, which is the second-deepest lake in North America (1,933 feet [589 m]).

A small volcanic island, named Wizard Island, is located in the western area of the lake. It is composed of a cinder cone that formed following the caldera's collapse. Today, the floor of the lake is the site of hydrothermal activity, a reminder to everyone that this volcano cannot be written off as extinct.

Opposite: Crater Lake, "shining jewel of the Cascades." In the foreground is Wizard Island, which formed 100 to 200 years after the collapse of the caldera.
Below: Geological map of the area of Crater Lake.

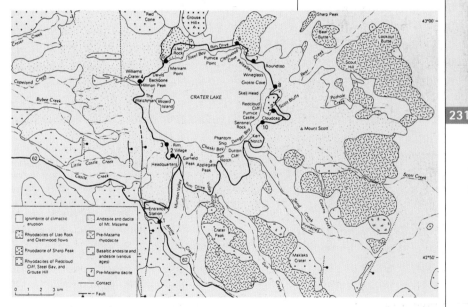

Access and principal attractions.

Crater Lake National Park is 93 miles (150 km) from Eugene, in southern Oregon. You can reach the lake from Eugene on routes 58, 97 and 138. From Medford, to the southwest of the park, take Highway 62 for a distance of 70 miles (113 km) to one of the entrances to the nature reserve. Hardly more than 62 miles (100 km) separate the lake from another spectacular site, Klamath Falls.

Special items of interest in the area include the Pumice Desert, to the north of the lake, the blue basin of the lake itself, the cinder cone of Wizard Island, which rises a little more than 655 feet (200 m) out of the water, and the Pinnacles, sand towers created by the erosion of fumarole deposits in pyroclastic flows that occurred during the enormous eruption that created the lake.

Continental margin

Stratovolcano

Prevalent volcanic activity. Lava flows, lava domes, Strombolian, Plinian, fumaroles.

General. With a volume that approaches 120 cubic miles (500 km³) (comparable to that of Japan's massive Fuji), Mount Shasta is the largest stratovolcano in the Cascade Mountains. Its peak, partially covered by a glacier, is 14,163 feet (4,317 m) high, rising 10,500 feet (3,200 m) over the surrounding plateau. The current edifice was formed atop the remains of a previous volcano that was active until two million to three million years ago. Alignments of cinder cones and small lava domes on the northern and southern slopes indicate the location of a fissure. A second fissure, facing east-west, produced the large parasitic cone on the western side of Shasta, called Shastina. Aside from Shastina, four principal craters are recognizable; the two youngest are less than 10,000 years old and date to a period during which there were repeated eruptions at intervals of between 600 and 800 years.

232

The zone to the northwest of the park, known as Lava Park, is striking for its complete absence of old trees. This absence results from the fact that most of the volcanic material from the volcano's last eruption, primarily pyroclastic flows and lahars, fell on this area. That eruption, Shasta's only historical eruption, took place in 1786 and was witnessed by the French explorer La Pérouse from a ship off the California coast.

Abundant solfatara activity takes place on the peak of the volcano.

Access and principal attractions.

Located 105 miles (170 km) south of Crater Lake in Oregon and 72 miles (115 km) northwest of Lassen Peak in California, Mount Shasta is the second highest peak in the Cascades (Rainier being the highest) and the highest in California. A little over 186 miles (300 km) north of Sacramento, following the route of the Shasta River, is Lake Shasta, located in Whiskeytown-Shasta-Trinity National Park. About 56 miles (90 km) north of the lake are small towns on the slopes of the volcano. To the northwest side, right at the foot of Shasta, is one of the largest and most impressive examples of an avalanche deposit, the result of the collapse of the prehistoric cone. The deposit is studded with the small hills known as hummocks.

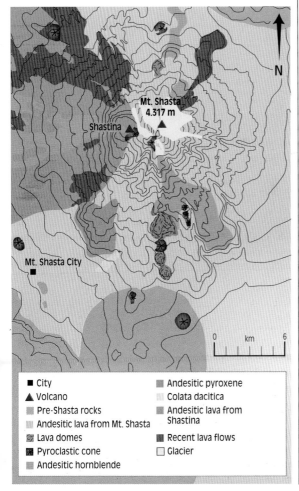

Mt. Shasta
4.317 m

Shastina

Mt. Shasta City

0   km   6

- ■ City
- ▲ Volcano
- Pre-Shasta rocks
- Andesitic lava from Mt. Shasta
- Lava domes
- Pyroclastic cone
- Andesitic hornblende
- Andesitic pyroxene
- Colata dacitica
- Andesitic lava from Shastina
- ■ Recent lava flows
- □ Glacier

Opposite top: Mount Shasta seen from the north. To the right is the cone of Shastina.
Opposite bottom: The cone of Shasta during the winter.
Left: Map indicating the locations of eruptive materials from the volcano. The lava flows covering the northwest slopes were produced by historical activity.

LASSEN PEAK, U.S.A.

Lat. 40.49 N – Long. 121.51 W – Alt. 10,456 feet (3,187 m) – 1203-08

59

234

Continental margin

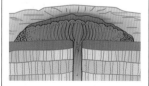

Lava dome

Prevalent volcanic activity. Lava domes, Pelean, phreatomagmatic, lahars.

General. Lassen Peak is composed of a lava dome with a base diameter of 1½ miles (2.5 km) that rises 2,885 feet (880 m) over pyroclastic deposits. Lassen is considered one of the most dangerous volcanoes in the Cascades Range. It is located in the southern area of the range, inside Lassen Volcanic National Park, and is the park's highest peak. The volcano sits atop lava flows discharged two million to four million years ago from a prehistoric eruptive vent known as Mount Tehama, located a little to the northeast, which disappeared more

than 350,000 years ago. Mount Lassen has a characteristic pyramidal form, truncated at the top, with steep slopes. Pinnacles of dacitic lava stick out from the slopes on the south and east sides. The principal eruptive phases of recent times include events, supported by dubious documentation, in 1650 and 1850–51.

The only documented eruptive events are those that took place on the summit of Lassen without interruption from 1914 to 1917.

In fact, from May 30, 1914, to March of the following year, Mount Lassen was the scene of some of the most violent eruptive activity known

Access and principal attractions. *Lassen Peak is about 125 miles (200 km) north of Sacramento. From Sacramento take Interstate 5 to its junction with 90, which you should take in the direction of the town of Chico. From there, travel to the junction on 70, which meets the turnoff for the park after a few miles. The attractions include Lassen Park Road, solfataras, the crest lakes, Bumpass Hell Nature Trail, Helen Lake, Lassen Peak Trail and the view from the peak of the volcano, including Mount Shasta and other members of the Cascades Range.*

on the continent of North America, including more than 150 explosive eruptions. The initial phase was considered hydromagmatic, caused by the interaction of molten magma with mud and snow that had accumulated on the peak during the previous winter. The lahars set off by these explosions devastated an area of about 4 square miles (10 km²). Successive explosions generated pyroclastic flows that descended the volcano's flanks to further damage the surrounding area.

In the years following those eruptions, activity at Lassen alternated between violent phases and phases that were more moderate.

There was very intense fumarolic activity until 1922, after which it dropped off dramatically, limited to just two sites, Bumpass Hell and Suspan's Springs.

Opposite: The lava dome of Lassen Peak.
Above: Lassen Peak during the summer. Lake Manzanita is visible to the right, Reflection Lake to the left.
Below: Geological map showing Lassen in relation to the prehistoric Mount Tehama.

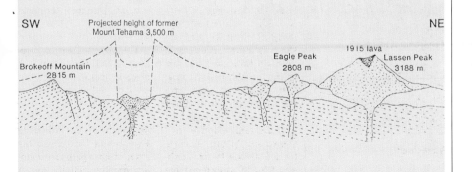

SW

Projected height of former
Mount Tehama 3,500 m

Brokeoff Mountain
2815 m

Eagle Peak
2808 m

1915 lava

Lassen Peak
3188 m

NE

236

Continental margin

Above: Mono Lake in the winter. The islands in the lake are rhyolitic lava domes.

Opposite: Geological map of the area with Long Valley Caldera, Mono-Inyo and Mono Lake.

Prevalent volcanic activity. Pelean, Plinian, Strombolian, phreatic, lava domes, lava flows.

General. The Mono-Inyo Craters, located in eastern California, are a chain of eruptive centers to the north of the large Long Valley Caldera. That caldera was formed 730,000 years ago following an immense eruption that ejected 144 cubic miles (600 km³) of pumice and ash. Today, those eruptive materials constitute the large ignimbrite deposit of the Bishop tuff.

The Mono-Inyo Craters extend over an area of about 31 miles (50 km), from the western edge of the Long Valley Caldera northward to Mono Lake. They are divided into three groups of structures called, from south to north, Inyo Craters, Mono Craters and Mono Lake. The local volcanic activity began with the eruption of fluid lava of basaltic composition but gradually moved to the north. Lava of rhyolitic composition began to be erupted around 35,000 years ago, first at Mono Craters and then at Inyo.

Mono Craters are an arching chain 10½ miles (17 km) long composed of roughly 30 domes, flows and craters that originated between 35,000 and 600 years ago. The most recent eruptions took place in the more northern area of the chain, from a 4-mile (6-km) long fissure that produced all together about 0.2 cubic miles (1 km³) of pumice, ash and obsidian flows.

Inyo Craters are a chain of eruptive vents 7½ miles (12 km) long that have created domes, flows and pyroclastic material of rhyolitic composition over a period of time between 6,000 and 600 years ago. Numerous phreatic explosions preceded and followed the most recent eruptive events, forming deep craters, two of which are Inyo Craters lakes. The most recent eruptive activity in the region was the eruption of a lava dome in the area of Mono Lake between 1720 and 1850, which led to the formation of Paoha Island.

Access and principal attractions. *The area of the Long Valley Caldera and Mono-Inyo Craters is easy to reach from San Francisco (217 miles [350 km]) by way of Yosemite National Park, the eastern border of which is just 12½ miles (20 km) from Mono Lake. The entire region has geological and natural attractions of enormous interest, including the many deposits from eruptions, such as the large ignimbrite deposits of Bishop tuff and the obsidian flows of Mono-Inyo Craters. There are also the famous granite peaks of Yosemite.*

237

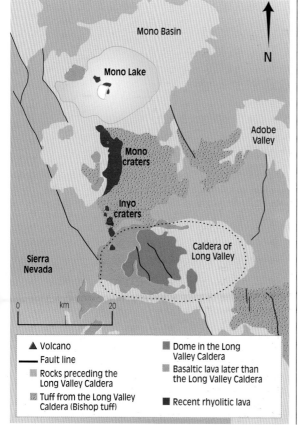

Mono Basin

Mono Lake

Mono craters

Inyo craters

Adobe Valley

Caldera of Long Valley

Sierra Nevada

0   km   20

N

▲ Volcano
— Fault line
▨ Rocks preceding the Long Valley Caldera
▨ Tuff from the Long Valley Caldera (Bishop tuff)
■ Dome in the Long Valley Caldera
▨ Basaltic lava later than the Long Valley Caldera
■ Recent rhyolitic lava

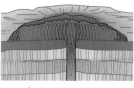

Lava dome

Cinder cone

YELLOWSTONE, U.S.A.

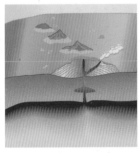

Hot spot

Caldera

Prevalent volcanic activity. Phreatic, thermal springs, geysers.

General. Yellowstone is probably the world's most famous volcanic area as well as America's most famous national park. It is also the first protected area in the United States, instituted in 1872. It is a broad plateau bordering the states of Wyoming, Idaho and Montana, with an altitude that varies between 6,890 and 7,875 feet (2,100 and 2,400 m), surrounded by mountains.

The volcanic activity at Yellowstone took place in three cycles over the course of the last two million years. Each cycle involved the emission of rhyolitic lava and was an explosive event of enormous size. These large explosive eruptions took place 2 million, 1.3 million and 600,000 years ago. The power of these events seems to have been at least 100 times that of the 1883 eruption at Krakatau in Indonesia. Following these large eruptive explosions a caldera depression formed in the Yellowstone area with an extension of 28 by 47 miles (45 by 76 km) and a depth of several hundred yards. Later volcanic activity, together with the erosive action of glaciers, partially covered and reshaped the walls of the caldera so much that now they are barely visible. The only activity in the area today involves hydrothermal displays, but these are exceptionally spectacular, fed by the high temperatures of the underground rocks.

In terms of the chemical composition of the water, three distinct thermal areas are recognized, one sulfurous, one alkaline and one calcium carbonate, which forms travertine deposits. There are at least 3,000 geysers in the area, concentrated in six main areas: Mammoth Hot Springs, Norris Geyser Basin, Lower Geyser Basin, Upper Geyser Basin, Shoshone Geyser Basin and the Heart lake Area.

Access and principal attractions. *Located in the far northwest of the state of Wyoming, Yellowstone National Park is easy to reach from all four points of the compass. One of the most famous natural parks in the world, Yellowstone offers such an enormous amount of geological and natural phenomena that several days are needed just to see them. From the volcanological point of view, the geysers are without doubt the main attraction. Old Faithful is the most* famous, followed by Excelsior, Great Fountain and Steamboat, the world's largest, able to shoot its column of steam and water 395 feet (120 m) in the air.

Opposite: Old Faithful, the most famous geyser in the park.
Below: Diagram showing the circulation of water in an aquifer and the hydrothermal displays of Yellowstone. The hydrothermal phenomena are kept active by the high temperature of the rocks and the infiltration of rain water.

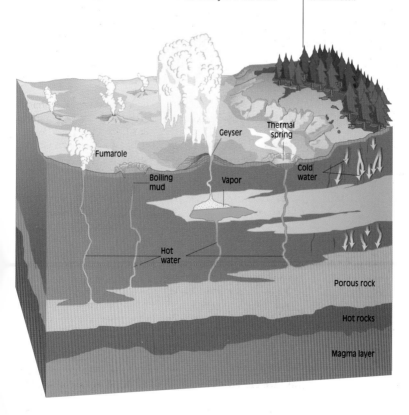

Fumarole

Geyser

Thermal spring

Boiling mud

Vapor

Cold water

Hot water

Porous rock

Hot rocks

Magma layer

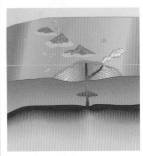

Hot spot

Shield volcano

Prevalent volcanic activity. Hawaiian, lava flows, lava lakes.

General. Kilauea is a shield volcano on the southeastern slope of Hawaii Island. It stands more than 4,000 feet (1,219 m) above sea level; to that height must be added another 12,140 feet (3,700 m) or so of the volcano's structure underwater, making Kilauea one of the largest active volcanoes on the planet.

The volcano was built up principally by lateral eruptions that took place along two rift zones that extend to the east and southwest from the summit caldera. This caldera is an oval depression 2½ miles (4 km) long and 2 miles (3.2 km) wide; its floor is covered by solidified lava, and it is surrounded by walls 394 feet (120 m) high. On the floor of the caldera is a circular depression called Halemaumau crater, which has been occupied many times by a lava lake. The crater's diameter has oscillated between 1,970 and 3,280 feet (600 and 1,000 m), with a depth that has reached 280 feet (85 m). Halemaumau appears in Polynesian mythology as the current home of Pele, the Hawaiian volcano goddess who travels from volcano to volcano along the arc of the Hawaiian Islands, bringing each to life with her arrival.

Along the east rift zone are numerous volcanic craters

along the Chain of Craters; these produced numerous eruptions that eventually centered on the crater called Pu`u O`o.

Because it is Hawaii's most active volcano in historical time, with enormous volcanic activity from the end of the 18th century to today, Kilauea is one of the most closely studied and monitored volcanoes in the world. It is the headquarters of one of the first volcanological observatories, the Hawaiian Volcano Observatory, established in 1912.

Access and principal attractions. *The caldera of Kilauea is easily reached by car. Many trails, varying in terms of difficulty, but not in terms of the spectacular nature of their views, leave from the Volcano Visitor Center located on the edge of the caldera: Crater Rim Trail, Halemaumau Trail, Kilauea Iki Trail, Chain of Craters Trail, Devastation Trail and Napau Trail. The awesome scenery and the many extraordinary geological marvels* include the Thurston Lava Tube, Pu`u O`o, and the current active crater, from which lava flows into the sea amid fantastic explosions and jets of steam.

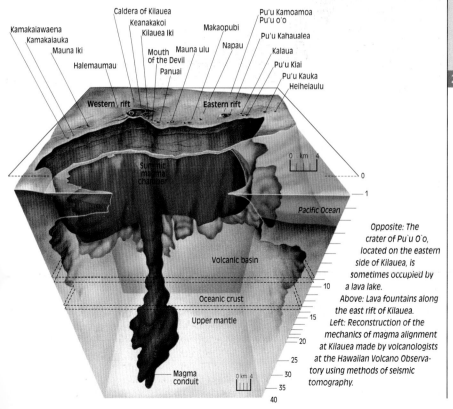

Kamakaiawaena
Kamakaiauka
Mauna Iki
Halemaumau
Caldera of Kilauea
Keanakakoi
Kilauea Iki
Mauna ulu
Mouth of the Devil
Panuai
Makaopubi
Napau
Pu'u Kamoamoa
Pu'u o'o
Pu'u Kahaualea
Kalaua
Pu'u Kiai
Pu'u Kauka
Heiheiaulu

Western rift
Eastern rift

Summit magma chamber

0 km 4

Pacific Ocean

Volcanic basin

Oceanic crust

Upper mantle

Magma conduit

0 km 4

0
1
10
15
20
25
30
35
40

Opposite: The crater of Pu`u O`o, located on the eastern side of Kilauea, is sometimes occupied by a lava lake.
Above: Lava fountains along the east rift of Kilauea.
Left: Reconstruction of the mechanics of magma alignment at Kilauea made by volcanologists at the Hawaiian Volcano Observatory using methods of seismic tomography.

241

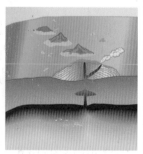

Hot spot

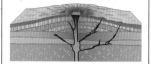

Shield volcano

Prevalent volcanic activity. Hawaiian, lava flows.

General. Mauna Loa is the largest active volcano on Earth. It stands almost 5½ miles (9 km) above the sea floor: there are 2½ miles (4 km) of its structure above sea level, with another 3 miles (5 km) underwater. If one adds the roughly 5 miles (8 km) of volcanic structure that, buried under their own weight, extend into the Pacific plate, the volcano's total height comes closer to 10½ miles (17 km), with a volume of more than 17,995 cubic miles (75,000 km³). The enormous mass of this volcano came into being through the accumulation of myriad highly fluid lava flows that traveled

dozens of miles from their point of emission before solidifying. On the summit of Mauna Loa is the Mokuaweoweo caldera, 3 miles (5 km) long and 1½ miles (2.4 km) wide, its walls up to 590 feet (180 m) high, which formed following a collapse of the summit area; it is flanked to the north and south by numerous smaller craters.

Mauna Loa is also one of the most active volcanoes on Earth, with more than 30 eruptions known to have occurred since 1843, the year in which historical record keeping of the volcano's activity began. Most of these eruptions were preceded by a series of earthquakes; usually, the magma is first erupted in the summit area, from inside the Mokuaweoweo caldera, with the production of enormous lava fountains that rise as much as 655 feet (200 m). The activity then moves to the system of rifts that spreads southwest and northeast from the summit. The lava flow that began during the 1859 eruption advanced for a period of 10 months, covering the 33 miles (53 km) between the eruptive vent and the coast and then continuing to flow underwater. In 1881 another enormous lava flow stopped just outside the capital city, Hilo, after covering a distance of 29 miles (47 km). In 1950 the volcano entered a period of quiet that ended in 1975, when it again began erupting from the summit, discharging 1.06 billion cubic feet (30 million m³) of lava between July 5 and 6. The last major eruption began on March 25, 1984, and ended April 15 of the same year.

Mauna Loa (together with its close neighbor Kilauea) is one of the most intensely studied volcanoes in the world. The ease of access and the excellent visibility of its stratigraphy make it the volcano with the best documented prehistoric activity on the planet. For this reason, models of its behavior can be evaluated against a period of activity that embraces roughly 10,000 years of eruptive history.

Access and principal attractions. *The trail that leads to the peak of Mauna Loa is one of the main attractions on the island of Hawaii. You have to be in good shape, however, for it is quite a demanding hike. The 18-mile (29-km) trail winds through Mauna Loa National Park. It's a day's walk to reach Red Hill refuge (9,845 feet [3,000 m]) and almost another full day to reach the Mokuaweoweo caldera at the top of the volcano, at an altitude of 13,681 feet (4,170 m). The walk along the endless lava fields and the view from the peak over the archipelago make for an unforgettable experience.*

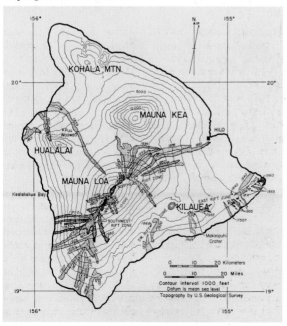

Opposite: The peak of Mauna Loa seen from southwest. Mokuaweoweo caldera and other smaller craters are visible along the rift system that extends to the southwest from the summit. In the background is Mauna Kea, another of Hawaii's recently active volcanoes, quiescent for about 3,600 years.
Left: Map of the island of Hawaii indicating the main historical lava flows of Mauna Loa, Kilauea and Hualalai, another volcanic center, which last erupted in 1800–01.

COLIMA, MEXICO

Lat. 19.51 N – Long. 103.62 W – Alt. 13,451 feet (4,100 m) – 1401-04

64

244

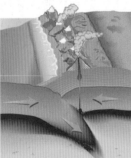

Continental margin

Stratovolcano

Prevalent volcanic activity. Plinian, Pelean, pyroclastic flows, lava flows, lava domes, fumaroles.

General. The Colima volcanic complex is composed of two stratovolcanoes, the older and by now extinct Nevado de Colima and the active Volcan de Fuego (or Volcan de Colima). About 10,000 years ago a gigantic explosive eruption destroyed the upper part of the volcanic edifice, forming a horseshoe-shaped depression with a diameter of 2½ to 3 miles (4 to 5 km) open toward the south. After this event, the volcano constructed a new cone inside the depression and entered a period in which extrusive and explo-

sive phases alternated. Written records of the volcano's activity began with the Spanish conquest of Mexico, and since that time the volcano has had a somewhat regular eruptive behavior in which four cycles have been recognized, each lasting about 100 years. Each cycle begins with the production of andesitic lava flows; these are followed by Plinian explosions accompanied by pyroclastic flows, and finally the creation and destruction of lava domes. The last explosive eruption began on January 20, 1913, when, without apparent warning signals, a Plinian column of gas and ash rose over the volcano. The ash was driven great distances by the wind, some of it being deposited on Guadalajara, 87 miles (140 km) to the north of Colima. Pyroclastic flows swept along the valleys, spreading destruction up to 9½ miles (15 km) from the crater. After the eruption, which lasted only four days, the volcano was quiet until 1957.

The activity that began in that year has continued to today and consists of the formation of lava domes inside the crater that then collapse, with lava flows that pour over the edge of the crater to flow down the steep flanks of the upper part of the cone, sometimes causing short, incandescent avalanches.

Access and principal attractions. Volcan de Colima is only 12½ miles (20 km) north of Colima, capital of the state of Colima in Mexico. Given the high amount of volcanic activity and the instability of the summit area, ascent to the peak is inadvisable except during particular periods of quiescence. We recommend that you obtain up-to-date information on the condition of the volcano from the authorities studying and monitoring it. The ascent offers views of recent deposits produced by Pelean-type pyroclastic flows, lava flows and the summit crater, occupied by a lava dome. The foot of the volcano is draped in a thick mantle of debris, deposits from avalanches produced by ancient cone collapses.

Opposite and below: Views of the summit of Colima. Lava flows that descend from the peak along the flanks of the cone are plainly visible. The current activity of Colima, combined with knowledge of its history, makes it seem highly likely that a large explosive eruption will occur within the next few decades. The consequences of such an event could be disastrous, since 200,000 people and the city of Colima are just 15½ miles (25 km) from the volcano.

PARÍCUTIN, MEXICO

Lat. 19.48 N – Long. 102.25 W – Alt. 10,400 feet (3,170 m) – 0101-01

Continental margin

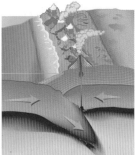

Cinder cone

Prevalent volcanic activity. Strombolian, lava flows.

General. Only rarely do volcanologists have the opportunity to witness the birth of a volcano. Parícutin, located in central Mexico and part of the Michoacán-Guanajuato volcanic field, which contains hundreds of cinder cones, provided this rare opportunity.

The story goes that Parícutin first appeared on February 20, 1943, in a former cornfield, a local peasant being the sole witness to this period of the event. The ground, shaken by earthquake tremors, rose several feet, and an old fissure opened farther to form a hole several feet wide out of which smoke, ash and sulfur rose, soon followed by blobs of incandescent scoria. This gave way to a phase of highly intense Strombolian activity. By 8 A.M. the next day a small cinder cone 33 feet (10 m) high had been formed; the eruptive activity increased so much that by noon the cone had reached 165 feet (50 m), and lava was flowing from its base. Covering ground at a speed of 3 miles (5 km) per hour, this lava flow soon reached and destroyed the peasant's farm. At the end of its first week of life, the cone had reached 460 feet (140 m) in height, and the level of Strombolian activity had not diminished in the least, with streams of molten lava being shot 3,280 feet (1,000 m) into the air. The most violent period of activity at Parícutin came between June and August of 1943. In October a secondary vent, named Sapichu, opened on the northeast side, destined to erupt in alternation with the main cone for several years to come. By the end of its first year of life Parícutin had reached a height of 1,102 feet (336 m). Over the next year andesitic lava flows discharged by new vents in the southeastern side of the volcano reached and destroyed the towns of San Juan de Parangaricutiro and Parícutin, the first 3 miles (5 km) from the crater, the other 1¼ miles (2 km). In the following years various eruptive vents around the volcano alternated in general activity, with Strombolian phases and large-scale emissions of lava. On February 25, 1952, after 9 years and 12 days, Parícutin ceased its activity, leaving a cinder cone 1,391 feet (424 m) high and an area of 10 square miles (25 km^2) completely covered by lava.

Opposite: Strombolian activity at Parícutin in a photograph from the early 1950s.
Below: Remains of the city of San Juan de Parangaricutiro. The church is one of the few buildings still visible above the lava.

Access and principal attractions.
Parícutin, located 199 miles (320 km) to the east of Mexico City, can be reached from that city by plane, train or car. The best base of operations for tours is Angahuan, 5 miles (8 km) to the northeast of the volcano. About 2 miles (3 km) to the south of Angahuan is the town of San Juan de Parangaricutiro, and another 1¼ miles (2 km) to the southwest is Parícutin, today completely buried by lava. Only a few of the town's taller structures, such as its church, rise above the blanket of lava.

THE AMERICAS

247

Continental margin

Stratovolcano

Prevalent volcanic activity. Plinian, Vulcanian, phreato-magmatic, lava flows, fumaroles.

General. Popocatépetl, nicknamed Popo, the most famous volcano in Mexico, has an almost symmetrical cone that rises majestically 13,780 feet (4,200 m) above the surrounding plain. It is about 37 miles (60 km) to the southeast of Mexico City. Its name, in Nahuatl, means "smoking mountain." This volcano, crowned by snow, is composed of two distinct structures. The old part, called Nexpayantla, ends in a rocky point (El Fraile), 12,467 feet (3,800 m)

high; the newer structure stands on the southeastern flank of the old Popo. The active cone is crowned by an oval crater measuring 1,310 by 1,970 feet (400 by 600 m) with vertical walls. On its floor is a small pyroclastic cone 115 feet (35 m) high surrounded by sulfur deposits from fumaroles.

Since the arrival of the Spanish in 1519, there have been 15 eruptions, and another two, dating to the years 1347 and 1354, are reported in Aztec documents. Most of the eruptive events of the last 600 years have not been exceptionally violent, involving the emission of gas and ash in clouds that reached heights of a few miles above the peak. However, geological studies reveal that in the period from 400 B.C. to A.D. 800 large explosive eruptions deposited

cinder and lapilli as far away as the area of Mexico City. In 1520, while besieging the ancient Aztec city that stood on the site of today's Mexico City, the Spanish conquistador Hernán Cortez sent an expedition to the top of the volcano to gather the sulfur needed to make gun powder for his cannon. The exploitation of the sulfur began at the beginning of the 20th century; in February 1919 explosives used on the floor of the crater caused a landslide that killed 17 workers.

In December 1994 Popo began activity again, causing the evacuation of about 50,000 people from the areas held to be in danger. There was further activity in May 2001.

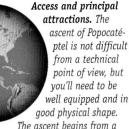

Access and principal attractions. The ascent of Popocatépetl is not difficult from a technical point of view, but you'll need to be well equipped and in good physical shape.

The ascent begins from a shelter located on the hill of Tlamacas, at 12,736 feet (3,882 m) of altitude, and it's a good idea to spend the night there to adjust to the altitude; furthermore, to make the climb you have to leave the shelter at 3 A.M. or 4 A.M. There are several trails to choose from. The most direct, but also the most difficult, winds around the eastern flank, passing by Teopixcalco (16,175 feet [4,930 m]), located under the peak of El Ventorrillo (16,404 feet [5,000 m]). The ascent takes nine to twelve hours; five to six are needed to get back down. Hiring a local guide is always a good rule to follow. The same goes for moving in a group, especially because the weather can change very quickly. There are plenty of travel agencies in Mexico City that offer tours and can furnish the necessary supplies. The major attractions are the summit crater, the panorama from the peak and the volcanic deposits along the trails.

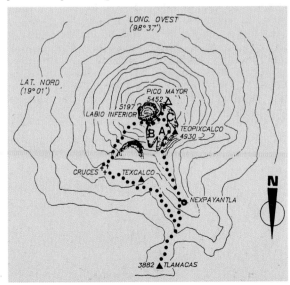

Opposite top: Intense fumarolic activity in the crater in November 1996.
Opposite bottom: The cone of Popo seen from the south, topped by a plume of gas and steam.
Left: Map of Popo indicating the principal trail to the peak (dotted line).

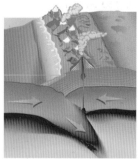

Continental margin

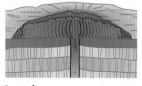

Lava dome

Prevalent volcanic activity. Plinian, Pelean, lava domes, fumaroles.

General. The volcanic center of El Chichón is isolated in the southeastern part of Mexico, in the province of Chiapas, halfway between the Trans-Mexican Volcanic Belt and the volcanic axis of Guatemala. In 1982, El Chichón produced the most important explosive eruption in Mexico during historical time.

Before 1982 the structure consisted of a lava dome covered by a dense forest with small fields of fumaroles at its summit. In March–April 1982, following a period of quiescence believed to have lasted 5,000 to 6,000 years, the vol-

cano began giving signs of coming awake with numerous earthquakes. The eruptive events began on March 29, with explosions that gave off a column of gas and ash. Forty minutes after the onset of the explosion, the plume of ash and gas had reached the stratosphere, expanding to form an enormous umbrella with a diameter of about 25 miles (40 km) and a height of 10½ miles (16.8 km). Two more large explosions of a violence similar to the first occurred on April 4. The ash-fall from part of the eruptive cloud rising from the crater caused the formation of torrents of incandescent ash and gas (pyroclastic flows and surges) that spread out at great speed from the eruptive vent, devastating an almost circular area of 59 square miles (153 km²), razing forests to the ground, and setting fire to nine villages. When the eruption finally ended, the ancient lava dome had completely disappeared, its place now occupied by a circular crater about 0.6 miles (1 km) in diameter.

The 1982 eruption killed 2,500 people, most of them victims of the pyroclastic flows. The high quantity of sulfur in the eruptive cloud caused the formation of tiny droplets of sulfuric acid in the stratosphere. Carried by the wind, these droplets of acid contaminated a 1,240-mile (2,000-km) wide band of the atmosphere that wrapped around the entire globe in 20 days. Some scientists believe that the eruption had a large impact on world climate, causing a drop in the average temperature of the Earth of 0.7°F to 0.9°F (0.4°C to 0.5°C) at the end of the winter of 1983.

Later studies of the 1982 eruption have indicated that El Chichón had in fact produced about 10 explosive eruptions similar in force to that of 1982 over the last 3,700 years. It is possible that some of these events had an effect on the development of the Mayan civilization, since most Mayan cities were located downwind of the volcano and thus would have been exposed to its abundant ashfall.

Access and principal attractions. *El Chichón is located 9¹/₂ miles (15 km) west of the village of Chapultenango, not far from the road that connects the capital of Chiapas, Tuxsla Gutierrez, to the capital of Tabasco, Villa Hermosa. Inside the summit crater is an acidic crater lake; there are also fields of fumaroles.*

Opposite: The crater that was formed by the 1982 eruption with its lake. The greenish blue color of the water results from its acidic nature. Fumarolic gases are constantly being fed into the water, and the coloration results from tiny altered materials held in suspension.
Below: Crater of 1982 seen from the south. The structure has a diameter of 0.6 miles (1 km) and a depth of about 820 feet (250 m).

Continental margin

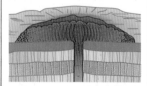

Lava dome

Prevalent volcanic activity. Pelean, lava domes, Plinian, fumaroles.

General. Santiaguito is a sort of lateral appendix to the cone of Santa María, an impressive stratovolcano located in western Guatemala in the area of Quetzaltenango. Its first historical eruption dates to 1902, when a gigantic Plinian-type explosion lasting 19 days emitted about 1.3 cubic miles (5.5 km³) of pumice and ash.

This left a crater in the southwestern flank of the cone of Santa María. On July 19, 1922, a vent opened inside the crater, beginning the creation of a lava dome, Santiaguito, which has produced intermittent eruptions ever since. The dome, made of very viscous dacitic lava, grew constantly and, on November 15, 1949, began to generate pyroclastic flows of the Pelean type. This highly destructive type of activity, particularly dangerous because it is so hard to predict, intensified in the period 1951–56 and has continued without interruption to today. In all this time it has proven quite difficult to even calculate the exact height of the edifice. In fact the dome of Santiaguito has no single well-defined peak, being constituted of a chaotic jumble of rocks, peaks and spires that forms an enormous promontory of volcanic material. On this massif are deep ravines out of which flow two streams, called Tambor and Nima Segundo. As many as 5,000 people may have been killed by Santiaguito's volcanic activity during the last century, but because of this very activity the volcano was studied in depth by volcanologists, increasing our understanding of the growth of lava domes and Pelean-type activity.

Access and principal attractions.
The volcano is located about 5¹/₂ miles (9 km) to the south of Quetzaltenango, and about 19 miles (30 km) to the west of Lake Atitlan. Two roads lead to the base of the Santiaguito lava dome. The first follows the access road to the Santa María volcano until the town of Mirador, where you descend a rocky gorge that ends at the base of the Santiaguito lava dome. The second road runs almost parallel to the first, to the west of it. A little before it begins the ascent of Santa María you leave the car and take a pathway to the right toward Tierra Colorada; from there, the second stop is Mirador de Magermans, once the site of a small mountain hotel. From there you head south to a fork in the road, at which point the true ascent begins. The path leads across the Valley de al Desolaciòn to reach a small flat space called La Isla, used for many years as a base camp by volcanologists. You can climb the lava dome, but there are dangers to consider. Up-to-date information on the state of the volcano is absolutely necessary, as is employing the services of an expert guide. The ascent offers the chance to study the phases of growth of a lava dome, with the extrusion of molten material and continuous small avalanches in the summit area. At times there are far larger collapses, attended by spectacular (and highly dangerous) Pelean-type pyroclastic flows.

Opposite top: The lava dome of Santiaguito seen from the east. The explosions that take place on the summit of the dome produce craters like the one visible at the center of the photo. Opposite bottom: The pyroclastic flow of 1989 began with the collapse of a portion of the Santiaguito dome. Volcanic gas and ash rose above the flow, generating a convective cloud. Below: Column of gas and ash from one of the explosions at Santiaguito in May 1992. These columns reached heights between 1,640 and 6,560 feet (500 and 2,000 m).

Continental margin

Stratovolcano

Prevalent volcanic activity. Vulcanian, pyroclastic flows, fumaroles.

General. Located 19 miles (30 km) west of Guatemala City, Fuego is considered the most active volcano in Central America, having erupted more than 60 times since 1524, the year of its first historical eruption. These events consist of Vulcanian eruptions from the central crater, often accompanied by pyroclastic flows; they have shown a tendency to cluster in groups over a period of 20 to 70 years with intervals of 80 to 170 years between groups. The last such group began in 1932 and is still in progress. The most recent large-scale eruption took place in 1974 and consisted of a series of Vulcanian explosions that caused an appreciable increase in the amount of solid particles in the stratosphere of the entire Northern Hemisphere. From September 1977 until mid-1979 Fuego was in a particular state of persistent activity during which columns of steam and volcanic ash were produced almost daily. This activity, relatively modest in terms of intensity, must be contrasted with the more than 25 single eruptions of greater intensity that Fuego has produced in the years since 1944.

Access and principal attractions. The ascent of Fuego presents no great difficulties from a technical point of view, but does require a certain amount of physical stamina since eight to ten hours are required to get up the mountain followed by another six to eight hours to get back down. The best idea is to make the trip in two steps, spending the night on the mountain. The best-known trail up leaves from Alotenango, 4¹/₃ miles (7 km) east of the volcano's peak; in that town guides and porters can be hired. At an altitude of 4,390 feet (1,338 m) you cross a bridge over the Rio Guacalate and take a gorge to the right to then cross a long stretch of lava fields deposited by the volcano. After a few hours the road turns toward the saddle

called Horqueta, between the cone of Fuego and that of another volcano on its edge, Acatenango; you then reach an open, rocky space. After another three hours of hiking you reach El Mirador, where you can camp for the night. The second part of the trip involves the final attack of the peak, which requires another five hours. There is no water on the volcano, so you must bring enough for at least two days.

Innumerable pyroclastic deposits, some of them accumulated over recent years, are highly visible and can be studied going up and again going down. There are fumaroles inside the crater summit, their plumes of

steam visible from quite a good distance. You must remember that Fuego is the most active volcano in Central America and that its eruptions are markedly explosive in nature, involving blocks shot into the air and pyroclastic flows down the volcano's flanks. It is therefore essential to obtain the most up-to-date information on the state of the volcano before undertaking the climb, and it is best to make use of expert guides. Observed from a safe distance, the volcano's explosive activity can be quite spectacular, with columns of gas and ash, pyroclastic flows and incandescent flashes above the crater.

Opposite and left: Views of the cone of Fuego and, behind it, its close neighbor Acatenango, another active volcano that gave off phreatic explosions as recently as 1926–27 and 1972.

PACAYA, GUATEMALA

Lat. 14.38 N – Long. 90.60 W – Alt. 8,373 feet (2,552 m) – 1402-11

70

256

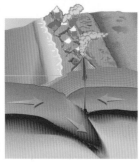

Continental margin

Compound volcano

Prevalent volcanic activity. Strombolian, Vulcanian, pyroclastic flows, lava flows, lava domes.

General. Located about 19 miles (30 km) to the south of Guatemala City, Pacaya is a compound-volcano massif built up in part on the rim of the caldera of Lake Amatitlàn, a depressed structure that is the site of the Calderas lagoon. Pacaya is composed of an ancient stratovolcano bearing a number of lava domes and by a younger stratovolcano generated by recent activity. The structure includes numerous peaks, including Cerro Chino to the northwest, Cerro Grande to the east and MacKenney cone to the south-west. MacKenney cone, the active peak of Pacaya, stands inside a large depression open to the southwest that was formed by the gigantic avalanche of an ancient cone. The shape of the summit part of the cone, which has several active eruptive vents, changes continually because of repeated explosive eruptions.

Pacaya has erupted at least 23 times since 1565, date of the first documented eruption. However, because it is located in an area that was hard to reach at that time, little is known of eruptions during the 16th century. Since 1962 the sinking of a summit portion of the volcano produced a depressed zone in which most recent activity has been concentrated, including the extrusion of a lava dome. Since 1965 Pacaya has been in a state of persistent activity, with alternating Strombolian and Vulcanian activity. Hundreds of Strombolian explosions have occurred every day, with streams of incandescent lava shot hundreds of feet into the air and the production of small lava flows.

On June 1, 1995, a plume of gas and ash 3¾ miles (6 km) high deposited ash up to 2½ miles (4 km) away; on May 20, 1998, for the first time in Pacaya's eruptive history, the volcano produced a rain of ash that fell as far away as Guatemala City.

Access and principal attractions. *Several routes lead up Pacaya, each of them requiring about an hour and a half to two hours of hiking. The most often used and best known of these trails begins at the level of San Francisco de Sales, on the northern side of the volcano. An hour's hike from this town brings you to the area known as La Meseta, from where you can reach the foot of the cone. From there you climb to the top.*

The volcano gets many visitors because of its perennial state of eruption, which includes jets of incandescent lava and extremely dramatic lava flows. The view from the summit is marvelous and includes many other active volcanoes in the area of Guatemala City. You must bear in mind that while Strombolian activity can be viewed from relatively close distances, there is always a modicum of risk involved, given the possibility of a larger explosion, which might include the ejection of lava blocks over a larger area than expected. Information on the current state of the volcano is always a necessity, as is trusting the advice of expert guides.

Opposite: Strombolian activity: the streams of incandescent lava shot into the air fall back around the crater; the glow visible to the right is from a lava flow spreading down the slope from the peak of the volcano. Left: Map of Pacaya indicating the major access paths and trails. The active cone, MacKenney, is inside the large depressed area (inside the curving line at the center). It was produced by an avalanche of the southwestern flank of an ancient cone. The volcano owes its name to the people of San Vincente Pacaya, as well as to the name of a species of edible palm that grows abundantly in the region.

CERRO NEGRO, NICARAGUA

Lat. 12.51 N – Long. 86.70 W – Alt. 2,215 feet (675 m) – 1404-07

Continental margin

Cinder cone

Prevalent volcanic activity. Strombolian, lava flows.

General. Located 13 miles (21 km) to the northeast of the city of León and 43 miles (70 km) to the northwest of the capital, Managua, Cerro Negro belongs to the Las Pilas–El Hoyo volcanic complex. It is part of an alignment of active volcanoes that runs northwest-southeast, thus parallel to the margin between the Cocos oceanic plate and the Caribbean continental plate. Other very active volcanoes in that alignment include Cosiguina, San Cristóbal, Telica, Momotombo and Masaya. Of these, Cerro Negro is the most active and also the youngest, having come into being in April 1850.

Since the time of its formation, Cerro Negro has erupted many times, with the emission of more than 10 lava flows. These flows, generated by a series of eruptive vents located along fissures at the base of the cone, are usually accompanied by Strombolian explosive activity. Some of these eruptions have been so explosive that they formed columns of gas, ash and lapilli reaching several miles in height and generating abundant rains of pyroclastic material over distances of more than 6 miles (10 km). Eruptions of this type took place in 1968, 1971 and 1992. During the 1992 eruption, plumes of ash and lapilli were formed that reached altitudes up to 5 miles (8 km), and blocks of lava and lava bombs up to 3¼ feet (1 m) in size were ejected from the volcano to fall around the crater within a range of 1,640 feet (500 m). Driven by the wind, ash fell as far as 12½ miles (20 km) away to the southwest. This eruptive phase began on April 9 and lasted until April 14, at which time the explosiveness diminished, and the columns of ash rose only 985 feet (300 m) over the crater. By April 15 the eruption had ended.

During the first three days of the eruption, its most violent phase, ashfall seriously damaged crops over an area of 77 square miles (200 km²), hitting many villages and even the city of Léon. The damage produced during this first phase was estimated at around $3 million; thanks to the timely evacuation of 10,000 people from the most endangered zones, there were no fatalities until April 5. By then, however, the accumulation of ash on roofs of buildings led to a number of collapses that took the lives of 50 people. Since 1992 the volcano has erupted in May–August 1995, with the production of a column of ash more than 0.6 miles (1 km) high and a small lava flow on the eastern side of the volcano, and again in August 1999.

Access and principal attractions. Many travel agencies in Managua offer assistance in organizing small expeditions to the volcano. Of course you can always go directly to Léon and engage the services of an expert guide. Cerro Negro is a splendid example of a cinder cone; its Strombolian activity, which occurs only at intervals, makes the view of the volcano even more interesting. There is also the city of León itself, capital of Nicaragua until 1858, which preserves much of its classical colonial style; some city neighborhoods are still particularly evocative of the past.

Opposite: At the center of the photo is the Cerro Negro cinder cone, seen from the southeast. In the background is Telica (3,314 feet [1,010 m]), and behind that San Cristóbal (5,725 feet [1,745 m]), both active volcanoes of the Nicaraguan volcanic axis.
Left: Vertical jet of scoria and incandescent ash during the 1971 eruption.

MOMOTOMBO, NICARAGUA

Lat. 12.42 N – Long. 86.54 W – Alt. 4,127 feet (1,258 m) – 1404-09

Continental margin

Stratovolcano

Prevalent volcanic activity. Strombolian, lava flows, fumaroles.

General. Momotombo is the southernmost active volcano of the Los Marrabios chain, in northern Nicaragua, part of the volcanic chain of Central America and location of some of the most active volcanoes on the planet. The birth of this chain is related to the subduction of the Cocos oceanic plate beneath the Caribbean continental plate, a movement that has generated an alignment of calc-alkaline volcanoes.

The cone of Momotombo is geometrically regular, with a summit crater open to the northeast. Although the pre-

historic activity of this volcano cannot be dated with certainty, it is believed to be young, meaning only a few thousand years old. Beginning in the 16th century its eruptions have followed one another with a rhythm of two per century, except during the 19th century, in which there were eight eruptions, one of which lasted fully eight years, from 1858 to 1866. The last eruption occurred in 1905: abundant flows of andesitic-basaltic lava were discharged from the summit crater, reaching the base of the cone. Since then the volcano has shown intermittent fumarolic activity. The level of this activity grew notably between 1973 and

1978, when the temperature of the fumaroles progressively increased from 482°F to 1382°F (250°C to 750°C). This drastic increase was probably related to an earthquake of medium-high intensity that occurred in March 1974 near the volcano. Intense fumarolic activity, with temperatures that reached spikes of 1580°F to 1652°F (860°C to 900°C) in 1985, continues inside the crater, at times making the walls of the crater incandescent.

Since 1982 there has been a geothermal power station generating 35 Mwatts at the southern base of the volcano, on the banks of Lake Managua.

Access and principal attractions. *A special permit is required to visit the Momotombo crater; it can be obtained from the Nicaraguan institute of energy in Managua or from the police in the city of León. The trail up the volcano can be reached only by off-road vehicle; the hike up the volcano is not overly difficult and takes about two or three hours. It begins on the lower flanks of the volcano, in forests that are home to several species of poisonous snakes; you then pass volcanic deposits along the steep slopes to reach the peak.*

Inside the crater are the high-temperature fumaroles, which sometimes make the surrounding rocks incandescent.
At the foot of Momotombo is an ancient volcanic depression with three small but splendid lagoons. The largest of these, Monte Galan, almost hidden by the forest, has a wonderful variety of fish and aquatic birds. It can be reached along the road from Hacienda San Cayetano, not far from the geothermal power station.

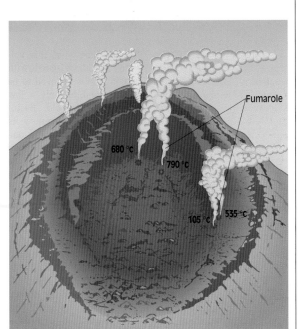

Fumarole

680 °C

790 °C

105 °C   535 °C

Opposite: The cone of Momotombo seen from the northeast. The lava flows visible in the foreground were discharged from the summit crater during the 1905 eruption. Behind the volcano spreads Lake Managua.
Left: Schematic drawing of the interior of the crater, indicating the location and temperature of the fumarolic emissions measured during a sampling carried out in 1985.
Above: Aerial view of the summit crater of Momotombo with smoke rising from the fumaroles.

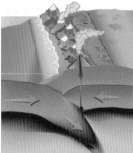

Continental margin

Caldera

Prevalent volcanic activity. Plinian and Strombolian, Hawaiian, pyroclastic flows, lava flows, lava lakes, fumaroles.

General. The caldera complex of Masaya lies along the Central American axis of active volcanoes, about 12½ miles (20 km) to the southeast of Managua. The caldera, 3¾ by 7½ miles (6 by 12 km) in size, contains numerous volcanic cones, the most important of which are San Pedro, Nindiri, San Fernando, Santiago (the current active cone) and Masaya. Although located in a geodynamic environment dominated by the convergence of the Cocos and Caribbean

plates and the production of calc-alkaline magma, the caldera of Masaya has basaltic magma of a tholeiitic composition. Equally unusual for basaltic magma is the explosive activity, with the formation of Plinian columns and pyroclastic flows, such as occurred in the eruption of 4550 B.C. This activity coexists in the Masaya caldera along with the effusive eruptions more typical of basaltic magmas. Because of the presence of the caldera, created following repeated collapses caused by the rapid emission of large quantities of magma, Masaya is considered a kind of basaltic equivalent of Indonesia's Krakatau. The Santiago crater occasionally fills with a lava lake, which is active for periods that run from weeks to years. In 1989

a lava lake about 165 feet (50 m) in diameter appeared following the collapse of the rocks that formed the floor of the crater. The lake formed 985 feet (300 m) below the rim of the crater; the crater's activity, including the formation of small lava fountains, indicated the process of degassing. This activity lasted only five weeks. The most recent activity began in June 1993 with the formation of a new lava lake, by now solidified, the surface of which was located about 165 feet (50 m) lower than the 1989 lake. The southern section of the caldera is occupied by a lake, Masaya Lagoon, which supplies drinking water to the city of Masaya.

Access and principal attractions. *A visit to the caldera of Masaya can begin at the entry to Masaya National Park, located 14 miles (23 km) from Managua and 6 miles (10 km) from the city of Masaya. A 4-mile (6-km) long asphalt road leads from the entrance to the border of the crater; short trails lead from there to various points of interest. Just 3 miles (5 km) to the south is a sandy beach along the shore of Masaya Lagoon, called Venecia, where you can go swimming. An 820-foot (250-m) long trail begins near the town of Nindiri and runs along the edge of the lagoon. Known as the Paseo de Hospital, this trail offers a fine panorama of the caldera and the many cones inside it. The caldera is home to an enormous colony of parrots.*

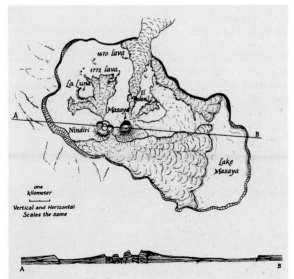

Opposite top: Dense fumarolic emissions from the Santiago crater, the source of recent activity in the Masaya caldera complex.
Opposite bottom: Aerial view of the Santiago cinder cone in 1968 with an active lava lake inside its crater.
Left: Map of the area of the Masaya caldera indicating the principal volcanic structures.

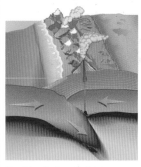

Continental margin

Compound volcano

Prevalent volcanic activity. Phreatic, phreatomagmatic, fumaroles, thermal springs, geysers.

General. Rincón de la Vieja, located in the north of Costa Rica, a few miles from its border with Nicaragua, is the most impressive volcanic complex in the Guanacaste Range. The volcano, the main attraction of a national park named for it, is composed of several eruptive vents aligned on a ridge running northwest-southeast: Braun (6,217 feet [1,895 m]), Von Seebach (6,106 feet [1,861 m]), Active Crater (5,906 feet [1,800 m]), Santa María (6,286 feet [1,916 m]) and another two smaller cones to the east. Recent activity has been concentrated inside Active Crater, which has a small lake. After a period of quiescence lasting more than 40 years (1922–66), the volcano returned to activity in the 1970s with a series of prevalently phreatic explosions, the last of which occurred on February 16, 1998. Some explosions have been phreatomagmatic, produced by the explosive interaction of magma inside the volcano with water from the surrounding rock. Some have involved columns of steam and ash several hundred feet high, with rains of mud along the slopes of the volcano and the formation of small lahars.

Access and principal attractions. From the city of Liberia, provincial capital of Guanacaste, you drive north a few miles along the Pan American Highway to Cerceda, where another road leads to the village of Curubandè. Northeast 0.6 miles (1 km) is the Hacienda Guachipelín, on the southern edge of the Rincón de la Vieja National Park, where you can spend the night and organize an excursion. Not far from the Hacienda Guachipelín is a forest ranger station that is the start of an easy 2-mile (3-km) trail through the forest that leads to one of the many hydrothermal spots in the area around the volcano. There you can see fumaroles, sulfur deposits and hot-water springs. Absolutely not to be missed is the view of the splendid fumarolic lagoon and the small mud volcano called

Volcancito. Another wonderfully evocative spot is right off the park's main pathway: Las Pailas, 8 acres (3.2 ha) of land immersed in the bright green tropical forest, studded with lakes of boiling mud, sulfuric vents, geysers and hot springs. The most difficult excursion in this area (5 to 5¹/₂ miles [8 to 9 km] of trail) is also one of the most beautiful in Costa Rica: the ascent of the volcano. While it presents no particular technical difficulties, it is long and tiring. Beginning at the forest ranger station there are two to three hours of hiking to reach the beautiful forest that covers the flanks of Rincón. The forest then suddenly ends and the hard part of the trail (one to two hours)

begins—the hike up the steep slopes of the volcano. What makes it worthwhile is the view from the peak, a panorama that takes one's breath away: Active Crater has a lake at its bottom, its color a luminous sky blue produced by the reflection of sunlight on the water, which contains particles of ash and chemicals from volcanic gas. To the west is the Pacific Ocean, and to the north the large Lake Nicaragua.

265

Opposite: Active Crater (5,906 feet [1,800 m]), center of recent eruptive activity.
Left: Aerial view of the volcanic complex of Rincón de la Vieja. Active Crater, near the center of the photo, can be recognized by its lake. The pale areas are those covered by ash during the November 1995 eruption. A local legend explains the volcano's name. A native princess named Curabanda, daughter of the ruler of these lands, fell in love with Mixcoac, chief of a neighboring enemy tribe. Her father, Curabande, learned of their love, captured Mixcoac, and threw him into the crater of the volcano. Curabanda, desperate with grief, hid herself on the side of the volcano and there gave birth to a son who she threw into the volcano so he could be near his father. She later became famous for her healing powers, and people named the place where she lived Rincón de la Vieja, meaning the "nook where the Old One lives."

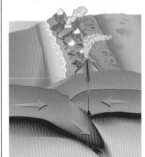

Continental margin

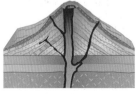

Stratovolcano

Prevalent volcanic activity. Strombolian, pyroclastic flows, lava flows, fumaroles.

General. Until the 1960s Arenal was not on the list of the world's active volcanoes. The last eruption had occurred around 1500, before the arrival of the Spanish, and in popular memory the volcano had never been active. The volcano's state changed drastically in July 1968. Three new craters aligned east-west were formed along the volcano's upper western flank. From the larger of these, at a height of about 3,610 feet (1,100 m), a series of explosions took place that threw gas, ash, scoria and blocks of magma into the air. This material rolled downhill, producing pyroclastic flows that killed 78 people in the village at the foot of the volcano. Large blocks were thrown up to 3 miles (5 km) away, forming impact craters with diameters as big as 200 feet (60 m). This eruptive phase lasted from July 29 to 31 and was followed by a second phase during which the prevalent activity was fumarolic, sporadically accompanied by the emission of volcanic ash. From September 14 to 19 the explosive activity returned, although with less intensity than before. On September 19 the emission of a flow of block lava began from the crater located at 3,610 feet (1,100 m).

Since 1968 Arenal has been in a perennial state of eruption; one of the most recent events was in April 2001. Strombolian eruptions take place on its peak at intervals of from a few minutes to a few hours, with plumes of gas and ash that in many cases rise more than 0.6 miles (1 km) in height.

There are frequent flows of block lava, which moves slowly. There are also occasional pyroclastic flows, which represent a constant danger to the inhabitants of the villages at the foot of the volcano.

Access and principal attractions. *From San José take National Road 1 northwest to the town of Sarchi Norte. From there take National Road 15 north; after four to five hours of driving you reach the small town of Fortuna, located 4 miles (6 km) to the west of the volcano. In the direction of Lake Arenal, 7½ miles (12 km) from the town, are the thermal springs of Tabacón. At night, while standing in the comfortably warm water of the springs, one can enjoy the spec-*

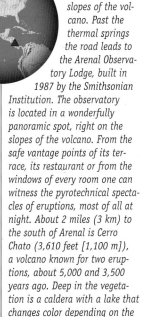

tacle of incandescent jets and lava flows on the summit and slopes of the volcano. Past the thermal springs the road leads to the Arenal Observatory Lodge, built in 1987 by the Smithsonian Institution. The observatory is located in a wonderfully panoramic spot, right on the slopes of the volcano. From the safe vantage points of its terrace, its restaurant or from the windows of every room one can witness the pyrotechnical spectacles of eruptions, most of all at night. About 2 miles (3 km) to the south of Arenal is Cerro Chato (3,610 feet [1,100 m]), a volcano known for two eruptions, about 5,000 and 3,500 years ago. Deep in the vegetation is a caldera with a lake that changes color depending on the time of day.

Opposite: Spectacular jets of incandescent scoria on the peak of the volcano. Still hot when they fall to the ground, they maintain much of their heat as they roll, leaving luminous trails behind them.
Left: Strombolian explosion at the peak of Arenal seen from the Arenal Observatory Lodge. In the foreground are the flanks of Cerro Chato.

POAS, COSTA RICA

Lat. 10.20 N – Long. 84.23 W – Alt. 8,885 feet (2,708 m) – 1405-04

76

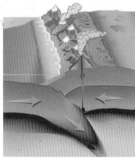

Continental margin

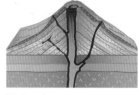

Stratovolcano

Prevalent volcanic activity. Phreatic, phreatomagmatic, Strombolian, fumaroles.

General. Located in Costa Rica's central volcanic chain, Poas is a stratovolcano topped by numerous cones and various craters, two of which have lakes. The volcanic massif is the heart of a national park of 5,599 acres (2,266 ha), the best equipped in Costa Rica. The summit craters and park are very popular with both Costa Ricans and foreign tourists. The main crater, 1 mile (1.5 km) wide and 985 feet (300 m) deep, contains one of the world's largest volcanic lakes. The crater periodically shoots up geyserlike jets of lake water,

the periods of inactivity between these events varying from a few minutes to weeks.

The first observations of eruptive activity at Poas date to 1828, and during the last 150 years the volcano has maintained a moderate level of fumarolic activity, associated with the presence of warm crater lagoons, interrupted by 40-odd eruptions, primarily phreatic or phreatomagmatic, but also Strombolian and Vulcanian. These eruptions have been concentrated in three principal periods of activity, from 1888 to 1895, 1903 to 1912 and 1952 to 1954. The most violent historical event was a phreatomagmatic eruption that took place on January 25, 1910.

Eruptive activity at the end of the 1980s was preceded by an increase in the number and intensity of seismic shocks in the area around the volcano and by a series of phreatic explosions in April 1988 that threw ash and mud out of the crater. In March 1989 numerous large fumaroles appeared inside the crater; by April the lake had been completely drained, while small sulfur lakes remained inside the crater, along with numerous cones of sulfur and mud.

The temporary draining of the lake led to the formation of a lake of liquid sulfur several feet in diameter, an unprecedented event on Earth. That same month saw strong eruptions of ash and mud rising up to 1.2 miles (2 km) to then fall back to the ground at distances of up to 0.6 miles (1 km) away.

It was necessary to close the park briefly. Today the crater gives off weak fumarolic activity with temperatures below 212°F (100°C). The water in the crater lake is highly acidic, with a pH less than 1.

Opposite: Panoramic view of the volcano's crater. It is 1 mile (1.5 km) wide and 985 feet (300 m) deep. Below: Corner of the crater lake of Poas, with fumarolic activity.

269

Access and principal attractions. *The first leg of the journey is from San José to Alajuela; 23 miles (37 km) north of Alajuela, along a road with many curves and hills to climb, is the entrance of Poas National Park. A short asphalt road leads from the entrance to a parking lot, from where you continue on foot. A 10-minute walk brings you to a pretty building housing a souvenir stand, small restaurant, volcanological museum with many interesting and instructive displays and a projection room where you can see films of the volcano's activity. A few minutes away is a belvedere overlooking the main crater. The early morning hours are the best time of day to enjoy the view, particularly in the dry season, before clouds obscure the panorama. The vision of the inside of the main crater is made even more dramatic because of its sheer size. Also very interesting is the nearby forest, often shrouded in fog, where you can see a large variety of tropical plants. A short trail through the forest brings you to another crater, inactive, with a pretty lake named Botos, surrounded by dense vegetation.*

IRAZÚ, COSTA RICA

Lat. 9.98 N – Long. 83.85 W – Alt. 11,260 feet (3,432 m) – 1405-06

77

270

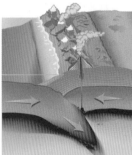

Continental margin

Stratovolcano

Prevalent volcanic activity. Vulcanian, fumaroles.

General. Irazú, the highest active volcano in Costa Rica, is located 12½ miles (20 km) northeast of San José and is part of the central chain of volcanoes. Irazú is also one of Costa Rica's most active volcanoes, holding an important record, with 20-odd eruptions since the first well-documented eruption in 1723. All these many eruptions have contributed to the great fertility of the soil in the Central Valley, world famous for the quality of its coffee production. The last large-scale eruption took place on March 19, 1963, a date made historical by the official visit on that very same day of U.S. President John F. Kennedy, a few months before he was assassinated. The eruption began an eruptive period that continued until February 1965. It produced moderate but uninterrupted rains of ash that damaged many coffee plantations, creating discomfort even in the capital. The ash washed away by rain accumulated in the bed of the Reventazón River, which rose, broke its banks, and flooded the valley, killing at least 20 people and destroying 400 homes and several farms.

At the peak of Irazú are two craters. The main crater, 3,445 feet (1,050 m) in diameter and

985 feet (300 m) deep, has a lake with green water; the other crater, named for the Spanish governor Diego de la Haya, the first person to keep precise records of the volcano's activity, is 2,264 feet (690 m) in diameter and 328 feet (100 m) deep. There are also two smaller craters, one of which contains a lake and a pyroclastic cone.

From January 1996 to June 1998 there were several seismic shocks, probably the result of the subterranean movement of magma. When the seismic activity began in 1996 the water in the lake was made turbid by the rise of volcanic gas.

Access and principal attractions. Located at the center of its own national park, Irazú can be reached by heading northeast out of the city of Cartago and taking the state road No. 8 (poorly indicated on road maps), which leads right to the site. At the end of an asphalt road is a parking lot and small ticket and information office. An easy trail takes you the roughly 0.6 miles (1 km) to the rim of the craters. This easy access to observation sites overlooking volcanic activity, in fact located above the main crater, makes Irazú an ideal volcano for everyone, offering the opportunity to

see a splendid example of double summit craters. From the top, on clear days, you see on the horizon both the Atlantic and Pacific oceans.

Opposite: The two summit craters of Irazú.
Below: View into Irazú's main summit crater with its emerald green lake.

RUIZ, COLOMBIA

Lat. 4.90 N – Long. 75.32 W – Alt. 17,457 feet (5,321 m) – 1501-02

78

272

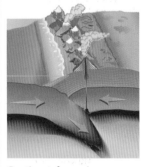

Continental margin

Stratovolcano

Prevalent volcanic activity. Mudflows, Plinian, Vulcanian, pyroclastic flows, fumaroles.

General. Nevado del Ruiz, which earned sad notoriety in 1985 for wiping away the city of Armero with a mudflow, is part of a volcanic chain made up of four main components, called, from south to north, Nevado del Tolima, Nevado Santa Isabel, Nevado del Ruiz and Cerro Bravo. The small massif of Nevado del Ruiz is composed principally of layered lava flows, with a relatively level summit covered by a 6½-square-mile (17-km²) glacier. The northern area of this summit area is occupied by the Arenas crater, in perennial fumarolic activity. On the

southwest slope, at a height of 16,076 feet (4,900 m), is a small parasitic cone named Olleta, its crater about 131 feet (40 m) deep.

The earliest recorded eruption dates to 1595, with another important eruption recorded in 1845; both were explosive, of relatively modest volume, but sufficient to produce mudflows caused by the melting of the summit glacier. The mudflow resulting from the 1845 eruption killed a thousand people in the Rio Magdalena valley.

In November 1984 the volcano gave signs of increasing fumarolic and seismic activity. After about one year of such warning signs, the eruption began on the afternoon of November 13, 1985.

The culmination of the eruptive phenomena was reached at 9:09 P.M. with the formation of a Plinian column about 19 miles (30 km) high that unleashed a rain of ash and lapilli to the northeast of the volcano. The Plinian phase was preceded and followed by the emission of clouds of ash and incandescent pumice (pyroclastic surges and flows) that rapidly melted the snow and ice of the glacier, releasing an enormous quantity of water (estimated at from 108 billion to 646 billion cubic feet [10 million to 60 million m³]). This water, mixed with detritus, rapidly formed a deadly avalanche (lahar) of mud and rock. The main mudflow swept down the Rio Lagunillas, reaching the city of Armero and razing it to the ground, killing 22,000 people. Other flows descended Rio Chinchiná and Rio Guali and caused terrible damage

to the two inhabited centers of Mariquita and Honda.

The volcanic activity had been foreseen, but it had proven impossible to avoid the terrible tragedy. In fact Ruiz presented a severe lesson in the difficulties of undertaking emergency measures in densely inhabited areas. It is a subject about which we still have much to learn.

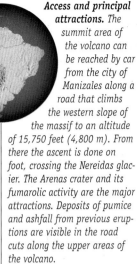

Access and principal attractions. The summit area of the volcano can be reached by car from the city of Manizales along a road that climbs the western slope of the massif to an altitude of 15,750 feet (4,800 m). From there the ascent is done on foot, crossing the Nereidas glacier. The Arenas crater and its fumarolic activity are the major attractions. Deposits of pumice and ashfall from previous eruptions are visible in the road cuts along the upper areas of the volcano.

Opposite: The summit area of the volcano seen from the west. The fumarolic activity is coming from the Arenas crater.
Left: Nevado del Ruiz seen from the northeast.
Below: The routes of the mudflows of November 13, 1985, and the locations of the towns that were struck in the valley of the Rio Magdalena (Armero, Mariquita and Honda).

273

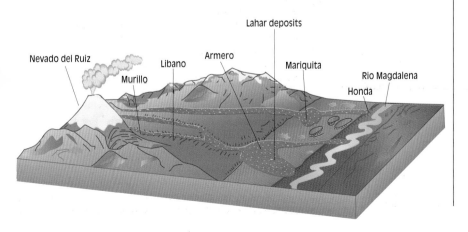

Nevado del Ruiz · Murillo · Libano · Armero · Lahar deposits · Mariquita · Honda · Rio Magdalena

GALERAS, COLOMBIA

Lat. 1.22 N – Long. 77.37 W – Alt. 14,029 feet (4,276 m) – 1501-08

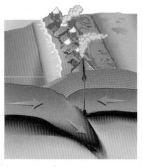

Continental margin

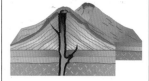

Compound volcano

Prevalent volcanic activity. Vulcanian, Strombolian, pyroclastic flows, lava flows, lava domes, fumaroles.

General. Galeras is located in Colombia's department of Nariño, 4⅓ miles (7 km) west of that department's capital, Pasto, in the southern part of the country. It is one of the centers of the volcanic axis that runs north from the border with Ecuador to Cerro Bravo. The main structure is a large stratovolcano with a base diameter of about 15 miles (25 km) and a volume of more than 72 cubic miles (300 km³). This stratovolcano stands inside a depression between the Cordillera Occidental and the Cordillera Oriental of the Andes.

An ancient avalanche truncated the summit area of the cone, leaving a large amphitheater-shaped depression open to the west. A new cone about 490 feet (150 m) high has formed inside the depression and has erupted various times during historical time.

The formation of Galeras began more than one million years ago; the avalanche that cut off the top of the cone occurred between 8,000 and 10,000 years ago. Most of the volcano's explosive eruptions have been modest in size, the outstanding exception being a large explosive eruption that took place between 4,100 and 4,500 years ago.

The first eruption of which there is written testimony dates to 1535, while the most recent occurred in January 1993. Numerous other eruptions have taken place over the course of the last four centuries; an eruption in 1866 produced the flow of lava blocks today called El Pedregal, which descended to the west, covering a distance of 3½ miles (5.5 km).

The volcano's return to activity in 1988 attracted the attention of volcanologists worldwide, and a conference was held in the city of Pasto to assess the dangers posed by the possibility of an eruption. On January 5, 1993, at 1:41 in the afternoon, while a group of volcanologists was taking samples from fumaroles near eruptive vents in order to evaluate the situation and make a prediction of future activity, the volcano suddenly exploded, taking the scientists completely by surprise.

The tremendous explosion killed six of the volcanologists (four Colombians, an Englishman and a Russian) and three journalists, and another three scientists were seriously wounded. The eruption ended June 7 of the same year.

Access and principal attractions. *The summit of the volcano can be reached by car from the city of Pasto, following a road that climbs the eastern slope of the mountain. The primary attractions are the amphitheater left by the avalanche, the active eruptive vents and the fumaroles.*

Opposite: The active cone inside the amphitheater-shaped depression, seen from the east.
Below: Corner of the active cone, seen from the south, during a period of heavy fumarolic activity.

REVENTADOR, ECUADOR

Lat. 0.08 S – Long. 77.66 W – Alt. 11,686 feet (3,562 m) – 1502-01

Continental margin

Stratovolcano

Prevalent volcanic activity. Strombolian, Vulcanian, lava flows.

General. The large volcanic cone of Reventador rises in the sub-Andean region that descends from the Cordillera Real toward Amazonia, about 56 miles (90 km) to the northeast of Quito. The volcano's eruptive history involves differing phases of activity during which the cone was alternately built up by the addition of erupted materials or partially destroyed by powerful

lateral avalanches or by the formation of summit calderas. Deposits from the avalanches are concentrated along the eastern flanks of the volcano but show up as far as 12½ miles (20 km) away. The current shape of the summit is a horseshoe open to the east inside which stands the active cone, which rises several hundred yards. Since 1541, the year of its earliest documented eruption, Reventador

of jets of gas and ejections of streams of incandescent lava. Three lava flows occurred during these years, involving the area between the current cone and the walls of the horseshoe-shaped depression. This activity ended toward the end of the 1970s. Because of its remote location and the relatively mild nature

Access and principal attractions. The volcano is located in El Oriente, the region of Ecuador east of the Andes. To reach the volcano take the Baeza–Lake Agrio road to the bridge over the Rio Reventador. There you park the car and make your way on foot into the tropical forest. The trail leads into a jungle, with stretches of dense vegetation alternating with open ridges from which you can catch magnificent views of the Amazon forest. About six hours of this hiking brings you to the bottom of the summit caldera; it is best to camp somewhere nearby for the night. The trail to follow the next day rises in a straight line toward the summit caldera. Then it snakes around the actual cone; another four to six hours of hiking are needed to reach the peak. The view from the peak of the volcano provides ample reward for the difficulty involved in reaching it; the setting is both desolate and beautiful, a place only the fortunate few ever have the chance to visit.

has produced another 24 Strombolian eruptions and at least four lava flows. Activity at this volcano became particularly intense around the middle of the 1970s, with Strombolian eruptions and periodic explosive discharges

of its eruptions, this volcano is not believed to present a major risk to the resident population.

Opposite top: The cone during its 1970s activity.
Opposite bottom: The volcano rises from the surrounding forest. Visible to both sides of the volcano are portions of the walls of the horseshoe-shaped depression in which the volcano stands (see page 74 for an aerial photograph of this area).
Left: The cone of Reventador in activity, discharging ash and gas.

PULULAGUA, ECUADOR

Lat. 0.04 N – Long. 78.46 W – Alt. 11,011 feet (3,356 m) – 1502-011

81

278

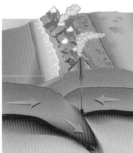

Continental margin

Caldera

Prevalent volcanic activity. Plinian, Pelean, lava domes, pyroclastic flows.

General. The volcanic center of Pululagua lies on the western edge of the inter-Andean depression, about 9½ miles (15 km) north of Quito. The combination of its proximity to the capital and the highly explosive nature of its eruptions makes it one of the potentially most dangerous volcanoes in Ecuador. It has the peculiarity of lacking a true volcanic structure. Unlike most of the other volcanoes in Ecuador, it is not a stratovolcano but rather a caldera depression, 2½ to 3 miles (4 to 5 km) in size and 985 feet (300 m) deep, inside which

stands the dacitic lava dome of Pondoña, 9,761 feet (2,975 m) high.

The volcanic activity of Pululagua can be divided in two periods: the earliest, of uncertain dating (less than 800,000 years ago), involved the formation of numerous lava domes arranged in two alignments running north-south about 2½ miles (4 km) apart. The formation of these domes was accompanied by avalanches of incandescent material that broke away from their flanks and filled the natural depressions. The more recent of the two periods begins with the major eruption of 500 B.C. (which is when the caldera was created) and runs up to historical time. The volcanic activity during this second period has involved highly explosive Plinian eruptions, the products of which, in particular notably thick layers of pumice and ash, cover the area around the caldera up to a distance of dozens of miles. The 500 B.C. eruption caused the sudden abandonment of several important pre-Columbian settlements to the south of the volcano in the northern areas of Quito (the Cotocallao civilization). Recently completed archaeological excavations have brought to light tools and earthenware buried by ash from the eruption and thus are exceptionally well preserved. The artifacts are today displayed in the archaeological museum of the central bank of Quito.

Access and principal attractions. *The first stop leaving from Quito is the so-called Mitad del Mundo, middle of the world, where, in 1736, a team of French geographers working on a geodetic survey established latitude zero, thus marking the halfway point between the two poles. By car, passing deposits of pumice and ash dating to the 500 B.C. eruption, you can reach the southern edge of the caldera (the Mirador terrace, at about 9,186 feet [2,800 m]). The splendid view from the terrace takes in the caldera and the central dome, Pondoña. An easy trail leads from Mirador down to the base of the caldera, about 985 feet (300 m). You cross the entire plain of the caldera, passing the small village of San Isidro and following the dome along its southeastern edge. Continuing to the north you leave the caldera to reach the confluence of the Rio Guayllabamba, about 2 miles (3 km) north of the caldera's rim. In all, this hike takes about three hours. The climb itself is about 1,640 feet (500 m).*

Because the trail is so easy, the landscape so attractive and the altitudes so low (less than 9,845 feet [3,000 m])—not to mention all the volcanological interest—this is an ideal itinerary for large groups.

THE AMERICAS

279

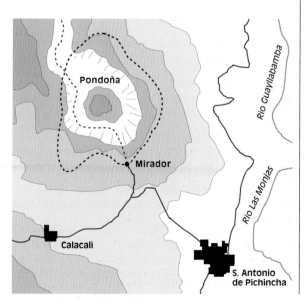

Opposite: The caldera of Pululagua, with the Pondoña lava dome at its center. There are many cultivated fields inside the caldera, the soil made fertile by the volcano.
Left: Map of the area of Pululagua indicating the roads (in red), the nearby towns, and the route that crosses the caldera from north to south (broken line). The hatchmarks indicate the edge of the caldera, which extends about 3 miles (5 km) along a north-south axis.

Pondoña

Mirador

Rio Guayllabamba

Rio Las Monjas

Calacali

S. Antonio
de Pichincha

Continental margin

Stratovolcano

Prevalent volcanic activity. Fumaroles, phreatic, Plinian, lava domes.

General. Pichincha, located immediately to the west of Quito, is composed of an ancient volcano, Rucu Pichincha, inactive for about one million years, and a more recent active cone, Guagua Pichincha, which has grown on the western slope of the old edifice. The active crater, with incessant, strong fumarolic activity, is inside a large horseshoe-shaped depression open to the west and 3¾ miles (6 km) wide. The major explosive eruptions at Guagua Pichincha of which there are written records took place between the 16th and 17th centuries. Put more precisely, they date to 1556, 1572, 1582 and 1660. Each of these events involved a powerful explosion, the formation of a Plinian column, and flows of ash, pumice and gas (pyroclastic flows and surges) that descended the western, uninhabited flank of the mountain. Studies of the prehistoric activity of the volcano indicate that other eruptions with similar characteristics took place around A.D. 550 and 970. During the 1660 eruption, the most violent during the last 2,000 years, the modern-day capital city of Quito, which at that time had only recently been captured by the Spanish, suffered significant damage from the ashfall, although most of the eruptive materials were directed to the west.

After the 1660 event the volcano went into repose except for small phreatic explosions in 1830–31, 1868–69 and 1881. After 100 years of complete quiescence this phreatic activity began again in 1981–82. Since August 1998 these explosions have occurred with greater violence and frequency, accompanied by earthquakes that have been registered on a seismic network. The phenomena could certainly lead in to a new explosion a short time.

Access and principal attractions. *The trip up the active crater of Guagua Pichincha begins in the town of Lloa, about 9½ miles (15 km) by stone road from the southern area of Quito. From Lloa a road climbs the mountain to the Civil Defense Refuge (15,256 feet [4,650 m]). This was set up for the use of scientists checking on the activity of the volcano, but it is also open to the public. A trail leads from the refuge to the northwestern border of the large horseshoe-shaped depression; all in all about a 20-minute walk. From there another 25 minutes brings you to the peak (15,696 feet [4,784 m]), which offers a panorama of the Andes and its numerous volcanoes, as well as a view of the active crater, located 3,280 feet (1,000 m) below. The climb down to this crater is long and difficult, both because of the drop (2,300 feet [700 m] from the edge of the depression), and because the trail is hard to follow and very easy to lose. Fumarolic and phreatic activity make the visit to the crater risky. The active crater, with its internal vents produced by recent phreatic explosions, is interesting, as are the fields of fumaroles and the large thermal springs located to the north. Deposits of pumice from the last 2,000 years of activity are visible in the road cuts around the refuge.*

Opposite: The active crater of Guagua Pichincha seen from the east. Visible at the center is the lava dome created after the large 1660 eruption, with phreatic craters and fumaroles. Left: Aerial photograph of Guagua Pichincha. The large depression open to the west was created by the collapse of the ancient cone, an event that took place more than 40,000 years ago. The active crater is in the paler area inside the depression.

ANTISANA, ECUADOR

Lat. 0.48 S – Long. 78.14 W – Alt. 18,875 feet (5,753 m) – 1502-03

83

282

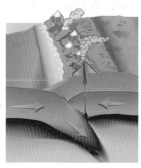

Continental margin

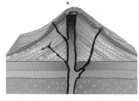

Stratovolcano

Prevalent volcanic activity. Lava flows.

General. Antisana, located in the eastern cordillera of the Ecuadorian Andes, about 31 miles (50 km) southeast of Quito, is the fourth highest peak in Ecuador. The large summit area, covered by ice and snow, includes four peaks that arose from the border of a single crater. The crater itself is completely occupied by snow and ice and only gives signs of weak fumarolic activity.

The eastern portion of the volcano, composed of an older edifice, is a summit truncated by numerous large, horse-shoe-shaped depressions, open toward the inaccessible re-

gions located to the northeast, south and east of the volcano. These depressions are the results of repeated large-scale avalanches of the ancient volcanic cone, presumably following major eruptions. The most recent cone, which makes up most of the part of the mountain that is visible from the area of Quito, stands on the northwestern part of the ancient structure.

No reliable information exists to indicate activity in the summit part of the structure during historical time, but at least two large eruptions have taken place from fissures that opened nearby. In 1728 an eruption of lava began from a vent that opened

10½ miles (17 km) to the west of Antisana; the resulting lava flow traveled about 7 miles (11 km) and was named Antisanilla. A second lava flow formed in 1773 from a fissure that opened about 6 miles (10 km) to the northwest of the volcano. This flow, known as Potrerillos or Papallacta, traveled 4¼ miles (7 km), along the way blocking the course of the Rio Papallacta and creating a lagoon.

Future eruptions on the summit of the volcano would present great danger since the heat of the eruption would melt large quantities of the snow and ice, leading to the formation of mudflows (lahars). These would be channeled into the valleys of the Papal-lacta, Quijos and Antisana rivers, endangering the inhabitants of the villages along those water courses. Equally at risk would be the Trans-Ecuador oil pipeline and the installations that collect, pump and transport drinking water to the capital city of Quito. The fall of pyroclastic material could hit the cities of Papallacta, Pifo, Tumbaco, Cummbayà, Pintag, the valleys of Los Chillos and, to a lesser degree, Quito itself.

Access and principal attractions.
Antisana is located along the Cordillera Real, the mountain range that runs along the eastern border of the Inter-Andean Depression, where Quito is located. This is a region of indescribable beauty, with the Andes gradually disappearing to the east in a diminishing series of hills leading toward the Amazonian plain. It is, however, a region without good roads. To reach the volcano from Quito you must first reach the Hacienda El Hato, to the southeast of the city. From there the approach to the peak takes you along a road and finally a trail. The 17th-century lava flows, Antisanilla and Papallacta, are fine examples of block lava of andesitic composition. The Papallacta lagoon, produced by the blocking of the river, is a site of great beauty and natural interest.

PACIFIC OCEAN

Pululagua
Reventador
Quito
G. Pichincha
Antisana
Cotopaxi
Sumaco
Quilotoa
Manta
Latacunga
Rio Napo
Chimborazo
Tungurahua
Riobamba
Sangay
Guayaquil

0 km   40

Opposite: Antisana, seen from the west, with its peak covered in snow. Left: Map of the central region of Ecuador, indicating the numerous active volcanoes in the area and highlighting (in orange) those treated in this book. The Pan American Highway, which crosses Ecuador longitudinally, is known as the Avenida de los Volcanes.

284

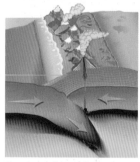

Continental margin

Stratovolcano

Prevalent volcanic activity. Pyroclastic flows, lahars, lava flows, fumaroles.

General. With its 19,393 feet (5,911 m) of height, Cotopaxi is one of the highest active volcanoes on Earth. It is known for the beauty of its regular cone, covered by a summit glacier that covers 3,280 feet (1,000 m) of height. On the summit is a crater 2,625 feet (800 m) in diameter and 1,095 feet (334 m) deep. Over the last 2,000 years Cotopaxi has had 17 major explosive eruptions. The eruption of 1877 caused the formation of *nuées ardentes* from the rapid melting of the glacier ice and the formation of enormous mudflows (lahars), which spread death and destruction along the valleys that run down the sides of the volcano up to a distance of more than 62 miles (100 km). Other large eruptions took place in 1534, 1742–43, 1744, 1766–68 and 1853, accompanied by mudflows and floods. Lava flows were produced in the last significant eruption, in 1903–04; the activity during recent years has involved periodic increases in fumarolic activity and the melting of a small area of ice in the summit area.

The mudflows have traveled in two principal directions, to the north and the area of the Los Chillos valley, where about 10,000 people live today, and to the south into the valley of Latacunga, home to 20,000 people and site of large agricultural and industrial zones.

The deposits from these lahars are often many feet thick, as in the case of the one that struck the San Gabriel textile mill in Latacunga, which was buried by the 1877 mudflow. In the Rio Cutuchi this lahar is estimated to have flowed at a maximum of about 1.8 million cubic feet (50,000 m³) per second, a power comparable to that of the largest rivers on Earth.

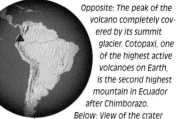

Opposite: The peak of the volcano completely covered by its summit glacier. Cotopaxi, one of the highest active volcanoes on Earth, is the second highest mountain in Ecuador after Chimborazo.

Below: View of the crater of Cotopaxi, located at 19,393 feet (5,911 m) of altitude.

Access and principal attractions. *From Quito you take the Pan American Highway about 25 miles (40 km) south; about 10 miles (16 km) past the town of Machachi, near the site of a NASA station, you turn off to the left, and after another 7½ miles (12 km) you reach the La Cienega farm, which is inside the park. From there it is necessary to take an off-road vehicle; an hour's drive reaches the José Ribas refuge, at an altitude of 15,750 feet (4,800 m), where it* is best to spend part of the night. Most climbers set off for the ascent of Cotopaxi just after midnight to avoid walking in the hottest hours of the afternoon, when the snow is softened by the sun, making the climb more difficult. The edge of the glacier is about an hour's hike from the refuge. From there, the peak of the volcano normally can be reached in four to eight hours of hiking. There are cracks and crevasses to watch out for in the glacier, but they are usually wide and thus quite visible. The climb does not involve any particular difficulties except for the high elevation. Ropes and crampons are absolutely necessary, most of all over the final 655 feet (200 m), which are steep. The ascent of Cotopaxi is one of the classics of mountain climbing in the Andes. The beauty of the natural scenery, the impressive view of the volcano's cone, and the summit glacier make the long climb well worth the effort. Once at the peak you can take a circuit of the summit crater, with its 2,625-foot (800-m) diameter, to take in the singular panorama of the Andes and the many volcanic peaks. Along the trail you encounter many types of volcanic deposits, including lava flows, pyroclastic flows, lahars and fields of blocks transported by flows.

Lat. 0.65 S – Long. 78.90 W – Alt. 12,841 feet (3,914 m) – 1502-06

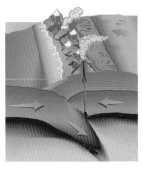

Continental margin

Caldera

Prevalent volcanic activity. Plinian, pyroclastic flows, fumaroles.

General. Located 22 miles (35 km) west of the city of Latacunga in the province of Cotopaxi, Quilotoa is one of the active volcanic centers in the Cordillera Occidental. It consists of a caldera depression 1½ miles (2.5 km) in diameter surrounded by steep walls cut by dacitic lava domes. Inside the caldera is a lake with a surface area of 1 square mile (3 km²).

The volcano's eruptive activity involves explosive eruptions with the formation of Plinian columns and pyroclastic flows that travel at high speed down the volcano's slopes. At least six such events appear to have taken place at Quilotoa during its relatively brief eruptive history, which began about 40,000 years ago. Deposits from the pyroclastic flows have accumulated in the valleys of the Zumbahua and Toachi rivers. The erosive action of the river water has shaped these deposits into highly unusual and evocative ravines and gorges, some of them 655 feet (200 m) deep.

The last eruption, which took place about 900 years ago, covered an area of 58 square miles (150 km²), with deposits accumulating in the Toachi River up to a distance of 15 miles (24 km) from the volcano. Phenomena related to modest eruptive activity are indicated during the years 1725 and 1740.

The surface of the lake in the caldera is at 11,485 feet (3,500 m) of altitude; its maximum depth is 740 feet (225 m). Since it is without outlets, changes in the level of the water result from rainfall and evaporation. Volcanic gas high in carbon dioxide is discharged into the lake. Some of this gas is constantly being released from the surface of the water, but much of it is accumulated in the water, forming gas pockets that are sometimes visible on the surface as spots of a paler color. If released, this gas could form a cloud of carbon dioxide that, descending the flanks of the volcano, would suffocate any animals or people in the area.

Opposite: The caldera of Quilotoa with its beautiful turquoise lake. Below left: Map of the area of Quilotoa. The red lines indicate the road that connects Zumbagua to Sigchos and the trail leading to the edge of the caldera.

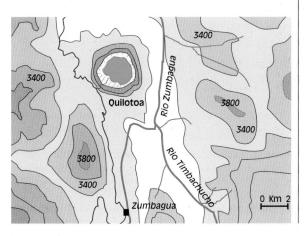

Access and principal attractions.

From the town of Zumbaqua, around 20 miles (32 km) west of Latacunga, head north on the road leading to Sigchos. The road runs about 6 miles (10 km) across an area covered with pyroclastic material produced during the last large eruption, about 900 years ago. At 12,566 feet (3,830 m) of altitude, leave the main road for a trail that runs about 0.6 miles (1 km) to reach the western edge of the caldera (12,598 feet [3,840 m]). The descent to the lagoon takes about half an hour, following an easy trail. The lake is surrounded by yellow, green and red incrustations produced by carbonic precipitation and iron oxides. There is a beach on the southwestern side of the lake used by the local inhabitants to water livestock.

Because of the lagoon's great beauty, the splendid, immense panorama and the relative ease of access, an excursion to Quilotoa is an ideal way to acclimatize oneself to hiking in the Andean altitudes before taking on more demanding ascents.

Lat. 1.47 S – Long. 78.44 W – Alt. 16,480 feet (5,023 m) – 1502-08

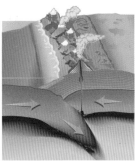

Continental margin

Stratovolcano

Prevalent volcanic activity. Plinian, pyroclastic flows, fumaroles, thermal springs.

General. Tungurahua, with the city of Baños at its northern base, is one of the most active volcanoes in Ecuador, having generated 20-odd eruptions in historical time. It has a regular conical shape; at its peak is a crater 600 feet (183 m) in diameter, inside which are many fumaroles that reach 172°F (78°C)—the temperature at which water boils at that altitude. The volcanic cone stands in the mountainous area of the Cordillera Oriental, and its lower slopes are exposed to constant erosion from the Chambo and Pastaza rivers. This erosion has produced spectacular gorges that are carved into deposits from lava and pyroclastic flows.

The volcano's record and shape indicate that a sudden eruption would involve the fall of ash and lapilli, the generation of pyroclastic flows and also lava flows. Such an eruption would threaten the nearby city of Baños as well as the dam on the Pastaza River immediately downstream of it.

The most devastating pyroclastic flows were emitted during the eruptions of 1773 and 1886 and were of a truly extraordinary size. These eruptions caused the melting of the small summit glacier with the consequent formation of mudflows. One of the secondary effects of eruptions of Tungurahua is that accumulated eruptive materials cause the temporary blocking of the water courses that run along its northern and western flanks; in the past, this blocking has led to the creation of temporary lakes that flooded the inhabited zones near the rivers. There are numerous thermal springs along the lower slopes of the volcano.

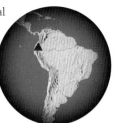

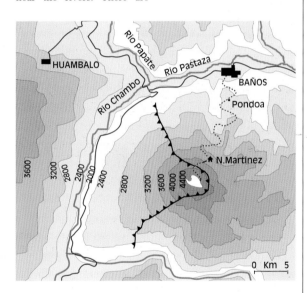

Opposite: The snow-covered cone of the volcano rises over the Andean plateau.
Left: Map of Tungurahua indicating the main access roads (in red) and the route leading to the top of the volcano (dotted line).

Access and principal attractions.

Although relatively easy from a technical point of view, the climb of Tungurahua requires a great deal of physical effort because the trail leads from an altitude of 5,906 feet (1,800 m) in the city of Baños up to 16,480 feet (5,023 m). From the city you reach the village of Pondoa then follow onward to the N. Martinez refuge at 12,467 feet (3,800 m). Some stretches of this trail take you through dense tropical forests of bamboo. The view of Chimborazo, the highest peak in Ecuador (20,561 feet [6,267 m]) located 25 miles (40 km) to the west, is impressive. The ascent of the peak begins at dawn, while the snow is still frozen. You hike in a southeasterly direction up to 13,123 feet (4,000 m) of height. Two hours of hiking brings you to an aluminum cross. From there the trails veers to the right, and after another two hours of hiking you reach the pair of summit craters (only the smaller is active). Be careful not to camp too near the fumaroles. Another 45 minutes of hiking across ice are needed to reach the highest part of the volcano; this stretch requires technical know-how and the use of crampons. The climb is thus long and difficult, but it is also fully repaid by the beauty and solitude of the locales you pass through, from the tropical jungle at the lower slopes to the ice-covered peak with its spectacular panorama of the Andean chain and Amazonia. It is also interesting in terms of volcanology and as a mountain-climbing experience.

SANGAY, ECUADOR

Lat. 2.03 S – Long 78.33 W – Alt. 17,159 feet (5,230 m) – 1502-09

87

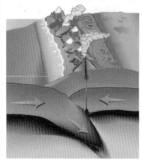

Continental margin

Stratovolcano

Prevalent volcanic activity. Strombolian

General. Sangay, located inside a national park in the province of Morona-Santiago (in the southeast of Ecuador), is the only Ecuadorian volcano in a state of persistent activity, and in fact it ranks among the world's most active volcanoes. Even so, because of its remote location is it practically unknown. In 1872, W. Reiss determined the height of the volcano to be 17,464 feet (5,323 m), and seven years later the famous English mountain climber Whimper reached the peak, which he later described as having been truncated by a large explosion. Moore, however, who climbed the volcano during a period of inactivity in August 1929, reported the cone intact. Today Sangay has a regular conical shape, its slopes angled at 35 degrees; the impressive structure rises from a height of about 5,250 feet (1,600 m) in the middle of the tropical forest to reach the extraordinary height of 17,159 feet (5,230 m). The summit is covered by a small snowfield. The northern slope of the cone has no plant growth and is instead covered by volcanic scoria given off during the continuous eruptions. A deep valley descends the eastern flank from the principal crater; it is most probably the primary channel for the *nuées* *ardentes.* An aerial overflight in 1976 brought to light the presence of three well-shaped summit craters. These craters are the site of typical Strombolian activity, with explosive periods that are repeated at regular intervals varying from a few minutes to a few hours, shooting jets of gas and fragments of incandescent lava many hundreds of feet into the air. This material then falls back onto and around the craters in a rain of fire. In August 1976 the members of a British scientific expedition, camping near the western crater, suffered serious damage as a result of the fall of such incandescent material. A state of uninterrupted activity is indicated from 1728 to 1916 and from 1934 to today.

Access and principal attractions. *The Sangay National Park is one of the most beautiful and wild in Ecuador. The best place to reach it from is the town of Alao, located in a fertile valley to the northeast of Riobamba and about two hours by car from that city. While in Alao food supplies should be bought to last the entire trip. The trail to the volcano leads out of the town and following three days of hiking up and down through dense Amazonian forest reaches the base of the volcano, where you must set up your base camp at 11,485 feet (3,500 m) altitude. From this safe distance you can observe the explosions of Sangay. Another day of climbing is needed to reach the top of the volcano; because of its persistent explosive activity getting too near the peak is obviously* very dangerous. In the company of an expert guide you can climb to a safe altitude from which the activity can be viewed close up; remember that the dangers include not just eruptive material but toxic gases and landslides. The return trip is done using the so-called Eten route, which involves a three-day hike to the town of Eten. From Eten you can reach the city of Baños by off-road vehicle. Sangay is one of the few volcanoes in the world in a state of persistent eruptive activity. The long approach hike through the Amazonian jungle is difficult but offers the chance to experience the area's wilderness beauty. The park is full of the vegetation typical of the Ecuadorian Sierras and is home to many animals, including mountain pumas, hummingbirds and llamas.

Opposite: The dramatic cone of Sangay towers more than 11,310 feet (3,600 m) over the Amazonian forest. Left: Aerial view of Sangay. The darker areas in the lower part of the cone are recently erupted material.

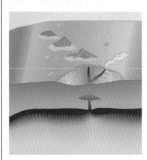

Hot spot

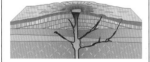

Shield volcano

Prevalent volcanic activity. Hawaiian, hydromagmatic, lava flows.

General. Fernandina, the most active of the Galápagos volcanoes, is also one of the most active volcanoes on Earth. The island is an enormous shield volcano with an area of about 19 miles (30 km) above sea level rising off the sea bottom from a depth of 3,280 feet (1,000 m). The shield structure is topped by an elliptical caldera depression 2.9 by 4.2 miles (4.6 by 6.7 km) in size and about 2,625 feet (800 m) deep. There is a lake inside the caldera. The caldera is surrounded by at least four curved fissures running concentric to the depression, the outermost of which, located about 0.6 miles (1 km) from the edge of the caldera, often produces lava, with the consequent formation of cinder cones. This system of fissures was probably produced by deformation accompanying the rise of magma. Another system of fissures extends around the caldera radially, in places intersecting the concentric fissures.

From 1813 to 1994 the volcano produced 23 eruptions, most of them effusive, occurring both inside the caldera and along the slopes of the volcanic structure. There have also been, although rarely, brief explosive eruptions inside the caldera, with jets of ash and steam, presumably hydromagmatic explosions resulting from magma coming into contact with the water in the caldera lake.

One of the most significant recent eruptions took place in 1968 when, following numerous earthquakes and small eruptions of lava along the eastern slopes of the volcano, the floor of the caldera dropped about 1,150 feet (350 m), making a large change in the topography of the caldera. The volume of the collapse was estimated at ½ cubic mile (2 km³), a far larger volume than the amount of lava erupted. Other eruptions took place in March and April 1984, and in September 1988 observation of the eruptive activity made during aerial overflights revealed the formation of lava flows along the floor of the caldera. These flows changed the shape and position of the caldera lake. A large explosive eruption in May 1991 caused the emission of a cloud of ash and steam that rose several miles in height.

The most recent eruption occurred in January and March 1994 when an eruptive fissure opened along the southwestern slope and produced numerous flows of a'a lava, some of which reached the coast, forming a delta structure about 1¼ miles (2 km) wide.

Access and principal attractions. *The Galápagos Islands are a protected area, and access is strictly limited. There are, however, various tourist agencies in Quito that organize guided tours of the islands, some of which include hikes up the caldera on Fernandina. The island is the site of a great variety of volcanic structures, including cinder cones. The summit caldera and its extensive fumaroles present a scene of striking beauty.*

Opposite: The foreground is taken up by recent lava flows; rising in the background is the unmistakable shield shape of the volcano of Fernandina.
Left: Map of the island of Fernandina. The summit caldera is 2.9 by 4.2 miles (4.6 by 6.7 km) with a depth of 2,625 feet (800 m).

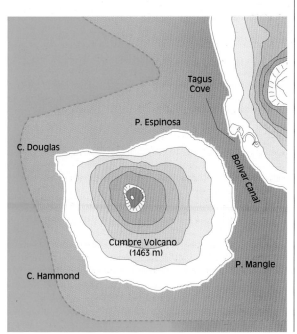

Tagus Cove

P. Espinosa

C. Douglas

Bolivar Canal

Cumbre Volcano (1463 m)

P. Mangle

C. Hammond

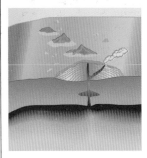

Hot spot

Shield volcano

Prevalent volcanic activity. Hawaiian, phreatomagmatic, lava flows.

General. Isla Isabela, the largest of the Galápagos Islands, is composed of seven shield volcanoes, each of which is crowned by a caldera, and five of which are active. The five active volcanoes are called, from north to south, Wolf, Darwin, Alcedo, Sierra Negra and Cerro Azul. The heights of the peaks vary from the 3,707 feet (1,130 m) of Alcedo to the 5,610 feet (1,710 m) of Wolf. The summit calderas of these volcanoes are primarily elliptical in shape and large in size: that on Sierra Negra measures 4⅓ by 5½ miles (7 by 9 km), but is only 360 feet (110 m) deep; the caldera on Alcedo is 3 by 4½ miles (5 by 7.5 km) and is around 850 feet (260 m) deep.

The seven shield volcanoes are composed almost entirely of highly fluid basaltic lava, emitted along eruptive fissures that spread out from the summits of the volcanoes or through arcing vents arranged around the caldera depressions. During eruptions small explosive activity occurs at the points where lava is emitted, leading to the formation of cinder cones. Alcedo is the only one of Isabela's volcanoes to have erupted white rhyolitic pumice and ash along with black lava.

Of Isabela's five active

volcanoes, Sierra Negra and Cerro Azul are those that have produced the greatest number of eruptions in historical time, each having erupted about 10 times since the early 1800s. The most recent eruption took place at Cerro Azul in the period of September–October 1998. The eruption began on September 15 and originated in vents both inside the caldera and on the slope of the volcano. The main lava flows were produced by a fissure that opened along the east-southeast slope. Over the course of the first 24 hours the lava spread 1¼ to 2 miles (2 to 3 km), moving first toward the east-southeast and then in the direction of the coast. An overflight of the volcano on October 1 revealed that the lava was coming from several vents, and that the shape of the lake in the bottom of the caldera had been partially changed. On that same day the flows descending the southwestern slope stopped their forward movement.

The eruption caused the mobilization of workers at the Charles Darwin naturalist station, who were called on to intervene when the lava flows threatened the nesting zones of Galápagos tortoises, a species already in danger of extinction.

Access and principal attractions. Many travel agencies in Quito offer tours to Isabela, including guides for the trip up the caldera. You can make the trip to the caldera of Alcedo on foot (a five-hour hike along the eastern coast) or by bus. The volcanoes of Isabela offer a great variety of lava structures, including tubes, cinder cones and tuff cones. The major source of interest, both volcanological and naturalistic, is probably the summit calderas with their extensive fumarolic fields.

Opposite: The Caleta Tagus on the western coast of Darwin, facing the island of Fernandina. The crater in the foreground is a fine example of a maar, produced by explosive activity following the interaction between seawater and magma.
Left: Map of the volcanic centers of Isabela and Fernandina.

Wolf
(1703 m)

Darwin
(1276 m)

Cumbre
(1463 m)

Alcedo
(1094 m)

Cerro Azul
(1689 m)

Santo Tomas
(1477 m)

SABANCAYA, PERU

Lat. 15.78 S – Long. 71.85 W – Alt. 19,577 feet (5,967 m) – 1504-003

90

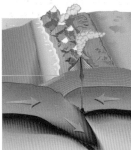

Continental margin

Stratovolcano

Prevalent volcanic activity. Strombolian, Vulcanian, Plinian, lava flows, lahars, fumaroles.

General. The volcanic massif of Sabancaya is located 50 miles (80 km) northwest of Arequipa, the second largest city in Peru, near the valley of the Colca River, home to about 30,000 people. The volcano rises from a 16,075-foot (4,900-m) high plateau to reach a height of a little under 20,000 feet. The summit area of the cone is covered by a glacier about 2 miles (3 km) in diameter. Sabancaya is one of a group of three stratovolcanoes (the other two are Nevado Hualca–Hualca and Nevado Ampato); the youngest of the group, it is located in the saddle between the other two. It is also the only one of the three to have erupted in historical time; its name means "tongue of fire" in the Quechua Indian language. The most recent lava flow, notable for its thickness, descended primarily along the northern slope of the volcano.

The most recent historical eruptions of which there are written records apparently date to 1750 and 1784. From the end of the 18th century to November, 1986 Sabancaya was quiescent, but in that month, following a gradual increase in fumarolic activity and earthquakes, the volcano returned to eruptive activity.

In December of that year a lava dome began forming inside the crater. This activity was accompanied by jets of incandescent scoria. During the first months there were also instances of Vulcanian-type eruptions, with strong explosions and jets of gray cinders reaching a height of around 4⅓ miles (7 km); the fall of this material covered an area of 12½ by 31 miles (20 by 50 km). Between June 22 and 25, 1988, the volcano gave off puffs of vapor every few minutes. On May 29 the exclusively fumarolic activity was interrupted by the return to explosive activity. The explosions, accompanied by dull rumbling, followed one another every 20 minutes, causing the formation of clouds of ash and gas that spread the smell of sulfur around the volcano. The eruptive plumes carried on the wind caused rains of ash to the northeast of the volcano at distances of up to 6 miles (10 km) away. Hot ash from recent eruptions of Sabancaya fell on nearby Nevado Ampato, melting areas of its glacier covering, and thereby exposing the mummified body of an Incan girl, part of a human sacrifice performed about 500 years ago. The find was made in September 1995.

Access and principal attractions. In the city of Arequipa you can organize an expedition to the crater of Sabancaya and the peak of Ampato. There are a dozen travel agencies that organize guided tours when the volcano's eruptive state permits it. The panorama from the top of Ampato (20,663 feet [6,298 m]) includes a dramatic view of Sabancaya and its eruptive activity. The nearby canyon of the Colca River, one of the deepest in the world, also merits a visit; several organized excursions include spending the night inside the canyon.

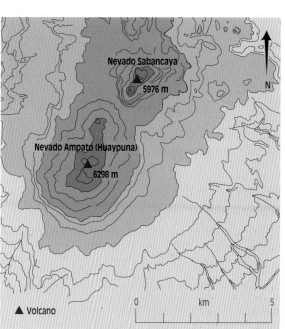

Opposite: The snowy cone of Sabancaya seen from the northeast with, behind it, the volcanic cone of Ampato, which is not known to have erupted during historical time.
Left: Topographic map of the Ampato–Sabancaya volcanic complex.

Nevado Sabancaya
5976 m
N
Nevado Ampato (Huaypuna)
6298 m
▲ Volcano
0   km   5

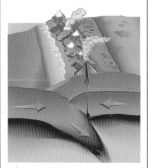

Continental margin

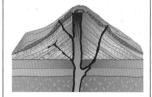

Stratovolcano

Prevalent volcanic activity. Strombolian, fumaroles.

General. This large stratovolcano towers over the city of Arequipa, 9½ miles (15 km) away. The summit has a crater 3,051 feet (930 m) in diameter, inside which is another crater, 1,804 feet (550 m) in diameter and bordered by layers of scoria. Inside it are many fumaroles. The volcanic structure is carved with deep channels, particularly in the northwestern area, while the northeastern flank is composed of a vast, arid zone covered by thick deposits of volcanic ash accumulated there by the wind.

Although the region of the volcano has been inhabited since pre-Columbian times, there is little record of its historical eruptions. The first eruption of which there is a record dates to the years between 1438 and 1471 and was handed down orally for about 150 years before being written down. While the volcano erupted, earthquakes shook the entire region; the explosions were heard hundreds of miles away, and the ancient city of Arequipa was annihilated, along with all of its inhabitants. Many other eruptions are reported as having occurred during that epoch, but the dates and the events themselves are highly uncertain. The most recent activity took place in May–October 1948, and it was probably in the course of that activity that the thick scoria deposits at the summit were formed as well as the lava dome inside the crater.

Access and principal attractions. *The ascent of El Misti presents no great difficulties outside those related to the high altitude. In fact, before making the climb you must first be acclimatized, which requires spending a few days at the departure altitude. Expert guides are needed, of course, along with ice axes, crampons and a good supply of water. There are two different ways to get up the volcano. The first and most difficult is the Apurimac trail, which takes two days and involves a change in altitude of 8,645 feet (2,635 m). You go by car from Arequipa to the base of the volcano, at an altitude of 10,500 feet (3,200 m); you then proceed on foot toward the peak, going along the edge of a large gorge and reaching, after seven hours of climbing and a 5,250-foot (1,600-m) change in altitude, the altitude of 15,750 feet (4,800 m), where you set up base camp. The next morning, after another four hours of hiking over ash deposits and overcoming a change in altitude of more than 2,950 feet (900 m), you reach the edge of the external crater at 18,700 feet (5,700 m) of altitude. From there a trail leads on another 30 minutes to the peak, indicated by a large iron cross. The descent is made very quickly (requiring just two and a half hours), using channels covered by volcanic ash. The second, far less demanding route begins with driving by car to the hydroelectric plant of Aguada Blanca at 11,810 feet (3,600 m) of altitude. From there five hours of hiking gets you to Monte Blanco (15,750 feet [4,800 m]), where you spend the night. The next morning, after another six hours of climbing, you reach the peak. The Andean Club in Arequipa organizes climbs, taking care of all the logistics, from the off-road vehicles to personal gear for the members of the group. Whichever route you choose, the climb is well worth the effort because of the nights spent on the mountain, the wonderful natural setting, the pyroclastic deposits, the summit crater, the many fumaroles and as always the astonishing panorama that is visible from the peak of the volcano.*

Opposite: The cone of El Misti rises over the surrounding plateau.
Left: Map of the southern area of El Misti indicating the ascent routes to the peak. As can be seen from the map, the city of Arequipa and its 600,000 inhabitants is quite close to the volcano. If the volcano returns to activity the city will be exposed to the real danger of pyroclastic flows and mudflows.

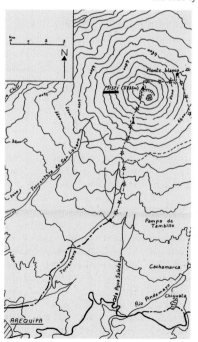

300

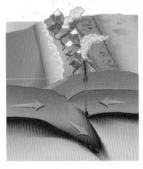

Continental margin

Stratovolcano

Prevalent volcanic activity. Phreatomagmatic, Plinian, Strombolian, Vulcanian, lava flows, fumaroles.

General. Lascar, located in the northern Chilean Andes, on the border of the Atacama desert, is an elliptical volcanic complex. It is topped by six partially overlapping craters aligned in a north-east direction. The two main craters, located at the far ends of the alignment, are 1 mile (1.6 km) apart. Each has a diameter of about 2,950 feet (900 m). The active crater, located at the center of the grouping, has a diameter of 2,625 feet (800 m) and a depth of 985 feet (300 m).

Lascar is considered one of the most active volcanoes in the Andes, best known for persistent fumarolic activity and periodic Vulcanian explosions. The edifice is the result of build-up from several eruptive cycles, which have produced pyroclastic materials and lava. The composition of the lava has varied over time. Numerous lava flows descend the northern and western slopes up to a distance of 4 miles (6 km) from the peak.

There have been at least 60 eruptive episodes over the last 150 years, many of them composed of phreatomagmatic explosions. Beginning in September 1986 the volcano produced a series of explosions that culminated in a large eruption in 1993. This event began on the evening of April 18 and reached its highest intensity the next day, with the formation of Plinian columns that reached heights of 14 miles (23 km) above the volcano. During the same day numerous pyroclastic flows were formed at the base of the columns.

The flows that descended the northern and northwestern slopes covered a distance of 4½ miles (7.5 km), damaging the town of Tumbre; those that descended the southeastern slope covered a distance of only 2 to 2½ miles (3 to 4 km). The April 19 explosive phase was followed by another, during the days of April 24 and 25, which involved Strombo-lian-type explosions and the extrusion of a lava dome on the floor of the crater. On April 26 a small explosion took place on the dome, after which the volume of the dome was reduced, probably because of the diminished internal pressure resulting from the liberation of gas. Small explosions of gas and cinders occurred in 1994 and 1995.

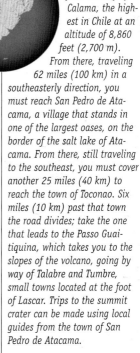

Access and principal attractions. *Lascar can be reached from the city of Calama, the highest in Chile at an altitude of 8,860 feet (2,700 m). From there, traveling 62 miles (100 km) in a southeasterly direction, you must reach San Pedro de Atacama, a village that stands in one of the largest oases, on the border of the salt lake of Atacama. From there, still traveling to the southeast, you must cover another 25 miles (40 km) to reach the town of Toconao. Six miles (10 km) past that town the road divides; take the one that leads to the Passo Guaitiquina, which takes you to the slopes of the volcano, going by way of Talabre and Tumbre, small towns located at the foot of Lascar. Trips to the summit crater can be made using local guides from the town of San Pedro de Atacama.*

Opposite: The volcano seen from the northwest with a large fumarolic plume rising over the summit.
Left: The interior of the active crater of Lascar with the lava dome, photographed in February 1989.

302

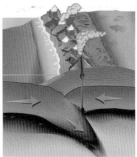

Continental margin

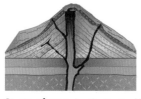

Stratovolcano

Prevalent volcanic activity. Strombolian, phreatomagmatic, lava flows.

General. Lonquimay is a small, flat-topped stratovolcano with a series of parasitic vents located along its base. The volcano formed near an area of structural weakness in the Earth's crust, a fissure zone known as the Cordon Fissural Oriental. The main cone is elliptical in shape and almost 2,300 feet (700 m) long at its main axis; the base of the entire edifice reaches 5 miles (8 km) in length. To the north are another three secondary eruptive vents with diameters of less than 985 feet (300 m), and on the southern side is another crater, whose diameter is more than 1,150 feet (350 m). In an area to the southwest of the main cone is a small chain of pyroclastic cones, 1¼ miles (2 km) long, from which a'a lava has been erupted many times.

The first documented historical eruption took place in 1853 with the formation of lava flows from the central crater accompanied by explosive phases that produced pyroclastic material. In 1887 a cycle of violent explosions began with the emission of blocks of lava from the central crater and from the area of the Cordon Fissural Oriental.

In 1933 effusive activity was noted at the volcano. The last eruption occurred in 1988.

The activity on that occasion was concentrated about 2.2 miles (3.5 km) from the main vent and led to the creation of a new crater, which was baptized Navidad since it opened on December 25.

The 18 days that preceded the eruption were marked by a series of earthquakes of growing intensity. This preeruptive phase ended in a phreatomagmatic eruption during which columns of ash and steam rose as high as 5½ miles (9 km) over the volcano. Wind action transported this volcanic ash to the east-southeast, reaching distances more than 373 miles (600 km) away. The phreatomagmatic phase was followed by a Strombolian phase, during which abundant lava flows were produced.

Although this eruptive activ-ity resulted in no human victims, it did a great deal of damage to agriculture and livestock. The magma erupted contained high amounts of gases like sulfur, chlorine and fluorine. The last of these in particular, absorbed into the soil through contact with volcanic cinders, led to the poisoning of much pasture-land, later causing the death of more than 10,000 head of cattle.

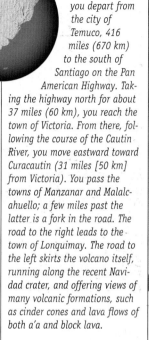

Access and principal attractions. To reach Lonquimay you depart from the city of Temuco, 416 miles (670 km) to the south of Santiago on the Pan American Highway. Taking the highway north for about 37 miles (60 km), you reach the town of Victoria. From there, following the course of the Cautín River, you move eastward toward Curacautín (31 miles [50 km] from Victoria). You pass the towns of Manzanar and Malalcahuello; a few miles past the latter is a fork in the road. The road to the right leads to the town of Lonquimay. The road to the left skirts the volcano itself, running along the recent Navidad crater, and offering views of many volcanic formations, such as cinder cones and lava flows of both a'a and block lava.

Opposite and left: Views of the snowy cones of Lonquimay. This stratovolcano has a 2,300-foot (700-m) long flat top and runs along the fissure zone called the Cordon Fissural Oriental.

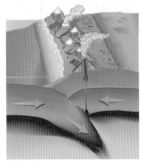

Continental margin

Stratovolcano

Prevalent volcanic activity. Strombolian, hydromagmatic, lava flows.

General. Llaima, one of the most impressive volcanic cones in Chile, stands about 8,200 feet (2,500 m) on the central-southern cordillera. Because of its height, most of its cone is covered by perennial snow. There are two active craters, one on the peak and the other to the southeast. The edifice was constructed atop an ancient caldera that was formed about 13,200 years ago and filled by the products of later activity. The eruption that created the caldera emitted a large pyroclastic flow, the deposits from which are known as Curacautìn Ignimbrite and the volume of which was around 5¾ cubic miles (24 km³). The flanks of the volcano are studded by 40-odd parasitic cones, most of them arranged in a northeast or south-east direction.

From 1600 to today the volcano has erupted about 30 times, most often of the effusive type, with the formation of lava flows and Strombolian explosions. The eruptions follow one another with a moderate degree of regularity and are separated by rest periods, but these have rarely exceeded 10 years over the last two centuries. The presence of ice and snow has often led to hydromagmatic explosions that have raised large columns of ash and steam. On October 15, 1979, the volcano began producing lava flows down its northeastern and western flanks, fed by the main crater. This eruption lasted all told 15 hours and was followed by a series of small phreatomagmatic explosions. During the last large eruption, on May reached a height of 2½ to 3 miles (4 to 5 km) in a short time.

Llaima presents an enormous potential danger because of the large masses of snow and ice accumulated on its flanks, for should these melt during an eruption, it would

Access and principal attractions. Setting off from the city of Temuco in the province of Cautín you head north along the Pan American Highway; after only about 5 miles (8 km) you leave the highway to travel east 27 miles (44 km) to reach the town of Cherquenco. It is there that the excursion to the volcano really begins. In Cherquenco you must engage a local guide to take you to the Llaima refuge at the base of the summit glacier, located at an altitude of approximately 6,560 feet (2,000 m). Along the route you'll see many lava fields as well as cinder cones, aligned along a system of fissures running northeast-southwest. Chilean volcanologists visited the site in March 1997 and reported abundant fumarolic activity in the northeastern area of the wall of the main crater. They also reported intermittent puffs of a bluish gas from the central crater. Before setting off for Llaima it is necessary to find out about the current state of the volcano, which has been known to give off frequent explosive eruptions of variable intensity.

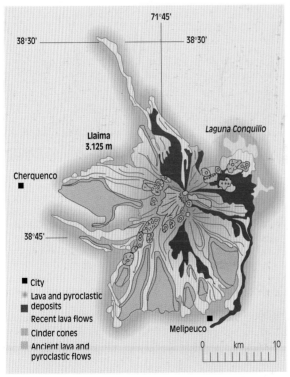

71°45'

38°30' — 38°30'

Llaima 3.125 m

Laguna Conquilio

Cherquenco ■

38°45' —

■ **City**

Lava and pyroclastic deposits

Recent lava flows

Cinder cones

Ancient lava and pyroclastic flows

Melipeuco ■

0 — km — 10

17, 1994, a 1,640-foot (500-m) long fissure opened along the northeastern slope of the cone and produced many lava flows. At the same time a column of gas and ash free large quantities of water that could form destructive lahars.

Opposite: The impressive cone of Llaima, one of Chile's largest and most active volcanoes.

Left: Geological diagram of Llaima indicating the areas covered by recent lava flows and the locations of the cinder cones along the flanks of the edifice.

306

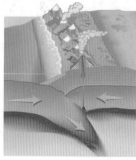

Continental margin

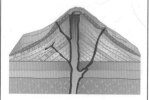

Stratovolcano

Prevalent volcanic activity. Strombolian, hydromagmatic, Hawaiian, lava flows, lahars.
General. Located in the southern Andean region of Chile, Villarrica has an eruptive pattern that is quite unusual compared to that of the other volcanoes in Chile. In fact it has been in a state of almost continuous activity for the past five centuries, primarily the effusion of fluid lava flows and the formation of fountains of basaltic lava. There are occasional explosions caused by magma coming into contact with water from the large glacier that covers the volcano.

The volcanic edifice is composed of a stratovolcano with two intersecting calderas in its central area; both calderas are elliptical and about 4 miles (6 km) long. The current volcanic cone is located inside the summit depression; it culminates in a crater about 1 mile (1.5 km) in diameter. Numerous parasitic cinder cones dot the flanks of the volcano.

Among recent eruptions, the one that occurred in December 1971 was among the most intense. The activity began at the end of October but reached its peak at 11:45

quence of this activity was the melting of ice atop the volcano with the formation of lahars that swept downhill at high speeds (up to 50 miles [80 km] per hour) and caused much damage and 15 deaths. The lahars exhausted their destructive force in Lake Villarrica to the north and Lake Calafquen to the southwest. There were also lava flows with an overall volume estimated at around 1 billion cubic feet (30 million m³).

Access and principal attractions. From the city of Valdivia you can drive to the base of Villarrica; the journey involves traveling northeast for about 75 miles (120 km). The volcanic activity is both spectacular and dangerous; it is essential to get updated information on the state of the volcano before undertaking any excursion. About 5 miles (8 km) away is the Las Cavernas system of lava tubes, which is open to guided tours.

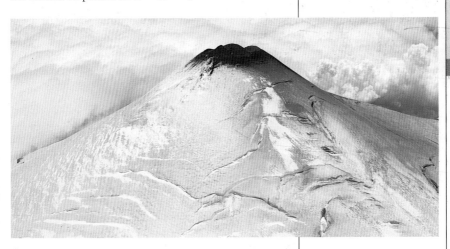

P.M. on December 29 when a fissure nearly 2½ miles (4 km) long opened along the central crater and gave off abundant lava, forming two fountains 1,310 feet (400 m) high. The fountains eventually took the characteristic shape of fiery butterflies with their wings spread. A conse-

Another phase of intense activity began in October 1984 with explosions at the summit, lava flows, avalanches of blocks of lava and snow, and the formation of a temporary lava lake inside the central crater.

Opposite top: Strombolian activity from the snowy cone of the volcano.
Opposite bottom: Aerial view of the summit crater, which measures about 1 mile (1.5 km) in diameter.
Above: View of Villarrica and its large summit glacier.

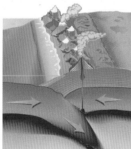

Continental margin

Stratovolcano

Prevalent volcanic activity. Plinian, phreatomagmatic, lahars, lava flows, fumaroles. **General.** Cerro Hudson is composed of a 6-mile (10-km) wide caldera located at the far southern end of the Chilean Andes. The volcano is located at the point where the oceanic Nazca plate is being subducted beneath the South American and Antarctic continental plates. The origin of the caldera is related to two large explosive eruptions that took place roughly 7,000 and 5,400 years ago. Two parasitic eruptive vents composed of cinder cones are located on the flanks of the volcano at a distance of 5 to 6 miles (8 to 10 km) from the caldera, one to the north and the other to the southwest.

The first historical eruption took place on August 12, 1971, when the caldera came to life with the formation of a Subplinian column that rose 7½ miles (12 km). The eruption melted some of the ice that covers the volcano, forming mudflows and lahars that ran down the slopes and killed 11 people. In 1973 the rise in temperature of the caldera formed yet another lahar that partially deviated the course of the Rio de los Huemeles. The last eruptive cycle was in 1991 and consisted of two explosive phases, the first on August 8 and 9, during which basaltic magma was emitted, and the second from August 12 to 15, when andesitic and dacitic magma was emitted. The first phase also involved a phreatomagmatic column 7½ miles (12 km) high, full of fluorine and chlorine. A 2½-mile (4-km) long fissure along the western edge of the caldera gave off lava flows. About three hours after the beginning of the eruption a lahar was formed that poured into the Rio de los Huemules, flowing over 31 miles (50 km) to reach the Pacific Ocean. During the second phase a Plinian eruption column rose 11 miles (18 km) high and, driven by winds, spread material as far as the Falkland Islands. The fluorine contained in the volcanic ash fell to the ground in rain, poisoning pasturelands in the Argentinean and Chilean Patagonia and as far away as Australia, eventually causing the deaths of thousands of head of cattle. Another lahar, larger than the first, devastated river valleys as far as the Pacific coastline.

Opposite: Spectacular aerial view of the caldera, including indications of fumarolic activity. The darker areas are where volcanic ash has been deposited on the glacier.
Below: Geological diagram of Cerro Hudson, indicating the path taken by lahars and the sites of the eruptive vents during the 1971 eruption and the two explosive phases of the 1991 cycle.

Access and principal attractions. *The remote area where the volcano is located is difficult to reach. One way is to leave from Argentina, beginning in the Atlantic Ocean seaport Comodoro Rivadavia. From there you head west to the Chilean city of Coihaique. A good road leads from there to Puerto Chacabuco, about 50 miles (80 km) away. There you can hire a guide for one of the various routes that lead to the slopes of the volcano. On the western side of the caldera are several rivers whose beds have been repeatedly invaded by lahars.*

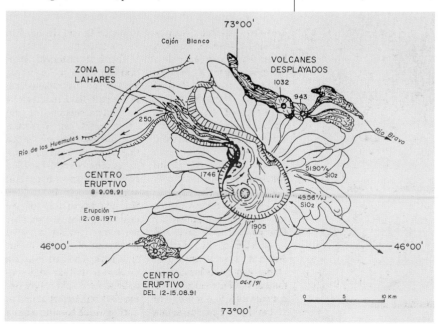

Island arc

Stratovolcano

Prevalent volcanic activity. Pelean, lava domes, fumaroles. **General.** Montserrat is one of the 122 islands that make up the island arc of the West Indies, which formed as a result of the subduction of the Atlantic plate beneath the Caribbean plate. The island, 5 by 10 miles (8 by 16 km) in size, has three main groups of hills called, from north to south, Silver Hills (1,322 feet [403 m]), Central Hills (2,428 feet [740 m]) and Soufrière Hills (3,002 feet [915 m]). This last group, on the southern end of the island, is the active volcano. Around 20,000 years ago, Soufrière Hills was the site of very intense eruptive activity, but there had

been no reported eruptions in historical time until 1995, and in fact many people believed the volcano was extinct. The only signs of activity had been important increases in seismicity reported in the years 1897–98, 1933–36 and 1966–67, but these had not ended in eruptions. Following a further increase in seismicity in the years 1992–94, the volcano returned to activity on July 18, 1995, with the opening of a fissure in a gorge between the old South Soufrière Hills (2,477 feet [755 m]) and Chanches Peak (3,002 feet [915 m]). The fissure gave off ash, steam and blocks of incandescent lava. By the end of September a lava dome had been extruded inside English's Crater. After the first signs of eruptive activity, 3,000 people left the island of their own accord; on August 21 the authorities ordered evacuation of the capital city, Plymouth, relocating thousands of people to the more secure northern end of the island. In November a further lava dome formed along the eruptive fissure that had opened on July 18. In the period March–April 1996 the first Pelean pyroclastic flows took place, beginning with incandescent avalanches along the flanks of the domes, the pyroclastic flows that were generated struck the southwestern area of the island, destroying several homes at Long Ground. At the same time an ash cloud rose up to 8½ miles (14 km), depositing 600,000 tons of ash on Montserrat. Other pyroclastic flows traveled down channels in the slopes of the volcano to strike the southern end of the island.

The activity of constructing and destroying lava domes is still going on today, and while

Access and principal attractions. *Access to the southern end of the island, site of the volcano, is still restricted by the state of alert, and just when the alert will end cannot be predicted. Since the late summer of 1998, however, it has been possible to visit the more northerly part of the area at risk. From there one can safely observe the volcano, its sole indication of activity being fumaroles at its summit.*

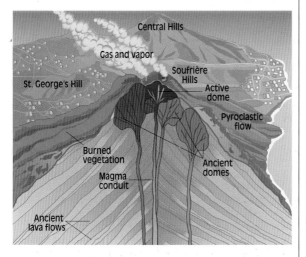

Central Hills
Gas and vapor
Soufrière Hills
Active dome
St. George's Hill
Pyroclastic flow
Burned vegetation
Ancient domes
Magma conduit
Ancient lava flows

material emptying to the east into the course of the Tar River. Between September 17 and 18 of that year numerous collapsed areas of the new dome led to a phase of paroxysmal explosions, during which a quarter of the volume of the dome itself was destroyed. The the level of eruptive activity is much less intense it is more than enough to prevent the return to normal life on the island. Thus this volcanic activity, the end of which cannot yet be foreseen, has had an enormous financial and social impact on the entire area.

Opposite: Pelean eruption on August 6, 1997. The explosion fractured part of the lava mass of the dome and threw the material into the atmosphere; the flanks of the dome collapsed under the weight of gravity, leading to the pyroclastic flows that spread along the flanks of the volcano, striking the surrounding valleys.
Left: Internal anatomy of the summit area of the volcano, with its numerous lava domes fed by volcanic conduits. In addition to the inhabited area of St. George's Hills there is the capital city of Plymouth, only 2½ miles (4 km) from the active lava dome.

Island arc

Stratovolcano

Prevalent volcanic activity. Pelean, lava domes, phreatic. **General.** The island of Guadeloupe is the largest of the Lesser Antilles, an island arc that formed following the process of subduction of the Atlantic plate under the Caribbean plate. The active volcano is in the southwestern part of the island and consists of a stratovolcano with a summit dome. In the course of its geological history Soufrière has produced numerous Pelean explosive events during which part of the summit dome collapsed, leading to debris avalanches that devastated the flanks of the volcano, until then covered by forests. Since the end of the 1600s the volcano has given off only phreatic activity, produced by the heating and explosion of water trapped inside the rocks of the volcano without the emission of new magma. Events of this type took place in 1797–99, 1809–12, 1837–38, 1903, 1956 and 1976–77. This last episode was preceded by seismic activity, including up to 1,000 events a day between July and August 1975, alarming the local population and the government of the island, in particular because of the tragedy that occurred in 1902 on nearby Martinique when the eruption of Pelée killed 30,000 people and destroyed the city of St. Pierre. The interpretation of the volcano's behavior, trying to determine whether the crisis would lead to a major eruption or would instead remain phreatic and thus of little danger, became the subject of dispute among international experts.

On July 8, 1976, at 8:55 A.M., a phreatic explosion took place accompanied by about 6,000 seismic shocks, the epicenter of which was the volcano.

The tense situation, further aggravated by the available means of communication, convinced the prefect of Guadeloupe that the experts who claimed an eruption was imminent were right. He ordered the evacuation of the island's 70,000 residents, an operation that began on August 15. The financial and social disruptions caused by this decision were enormous. From that July of 1976 to April of 1977 the volcano produced an almost uninterrupted series of phreatic explosions, but never a true

magmatic explosion, and the erupted material fell in the immediate area of the summit without damaging in any way the human structures on the island. This episode, the only large-scale false alarm to have occurred up to now, exacerbated the debate among volcanologists, civil authorities and the members of the mass media over how to handle similar situations. It also marked an important step in the history of volcanology, however, since the scientists had had the opportunity to come by a great deal of experience in the direction of volcano emergencies.

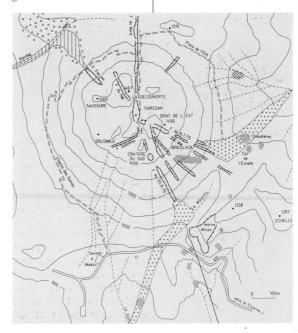

Access and principal attractions. *From the airport of Pointe-à-Pitre you travel to Basse-Terre, the western half of the island, where trips to the volcano and the lava dome can be organized.*

Opposite: In the foreground is the La Citerne crater, full of rain water; behind it to the left is the peak of L'Echelle, and past that is the lava dome.
Below: Map of the summit dome of La Soufrière de la Guadeloupe indicating the fissures that have opened (toothed black lines), the lahar deposits of 1976 (areas filled with small dots), the lahar deposits of 1837 (vertical lines) and the areas with fumaroles (filled with tiny squares).

PELÉE, MARTINIQUE, WEST INDIES

Lat. 14.82 N – Long. 61.17 W – Alt. 4,583 feet (1,397 m) – 1600-12

99

314

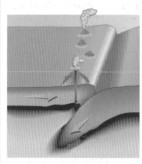

Island arc

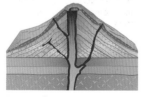

Stratovolcano

Prevalent volcanic activity. Pelean, Vulcanian, lava domes, fumaroles, phreatic.

General. Mount Pelée is composed of a stratovolcano that emits andesitic magma. This magma is too viscous to produce lava flows and instead forms lava domes at the summit of the volcano. The growth of these domes is often interrupted by avalanches or explosions that produce quantities of gas, ash and incandescent lava fragments that then flow along the many ravines running down the eroded flanks of the volcano to reach the coast of the island.

The eruptive cycle of 1902–05 followed a period of rest that had lasted about three centuries, interrupted in 1792 by two small phreatic explosions and in 1851–52 by some weak explosions that were probably phreatomagmatic.

Early in 1902 activity began again with an increase in fumarolic activity at the summit crater, Etang Sec, numerous slight earthquakes, and phreatic activity beginning on April 23. On May 2–3 the activity intensified, assuming a phreatomagmatic character. The accumulated erupted material, moved by rain, produced a series of lahars that killed more than 400 people at the Guérin factory in the village of Precheur. On May 8 at 8:02 the summit dome exploded, leading to the formation of a *nuée ardente* that moved south at a speed of more than 93 miles (150 km) per hour and razed to the ground the city of St. Pierre, only 5 miles (8 km) from the crater, killing all of its 28,000 inhabitants except one and destroying 16 ships anchored in the port. Other highly intense events took place on May 20 and August 30, causing another 1,000 deaths in the village of Morne Rouge. Less violent explosions took place on May 26, June 6, July 9 and from the end of 1902 until July 1905, generating low-velocity pyroclastic flows.

The 1902 eruption of Pelée gave the name to a particular category of explosive vol-canic activity, called Pelean and charac-terized by the growth and de-struction of lava domes with the formation of pyro-clastic flows. On March 30, 1903, at the end of the first erup-tive phase, a new lava dome, 1,000 feet (305 m) high, rose from the volcano's crater. The last eruptive cycle, which extended from 1929 to 1932, was also characterized by Pelean activity. Because of the experience acquired so sadly at the beginning of the century, the villages were rapidly evacuated on this occasion, avoiding the loss of human life.

Access and principal attractions. From the city of St. Pierre, rebuilt after the 1902 eruption, various paths lead north-ward across the deep Rivière Blanche gorge to climb to the peak of Pelée. The ascent of the moun-tain offers the opportunity to observe at close hand the vol-canic materials deposited by the many nuées ardentes, and there are also the lava domes on the summit. You are well advised to visit the volcanology museum in St. Pierre, which has many dis-plays relating to the 1902 eruption and the destruction of the city.

Opposite: Mount Pelée seen from the south. The old city of St. Pierre, razed to the ground by the 1902 eruption, stood to the left, just out-side the area of this photograph. That event gave the name to a type of volcanic activity, today called Pelean, which involves the growth and destruction of lava domes with the consequent formation of pyro-clastic flows.
Left: Map showing the areas destroyed by the pyroclastic flows during the eruption of May 8 (gray area) and August 30, 1902 (diagonal lines), during which 28,000 people lost their lives in the city of St. Pierre and 1,000 in the village of Marne Rouge.

316

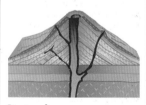

Island arc

Stratovolcano

Prevalent volcanic activity. Phreatomagmatic, Pelean, lava domes, fumaroles.

General. This stratovolcano is located in the northern area of the island of St. Vincent. Its activity is quite similar to that of Pelée, involving the periodic construction and destruction of lava domes with the generation of dangerous pyroclastic flows. The flanks of this volcano are cut by deep ravines that have become the channels used by the pyroclastic flows. At the summit of the volcano is a large crater, 1 mile (1.6 km) in diameter, located in the southern area of an older crater 1¼ miles (2 km) in diameter and open to the southwest as a result of slope failure. Inside the active crater is a lake with an island composed of a lava dome that came into existence following the eruptions of 1971–72 and 1979.

Since the 18th century there have been seven eruptions, the most violent being those of 1718, 1812, 1902–03 (18 days before the famous eruption of Pelée) and 1979. The 1902 activity produced pyroclastic flows that devastated the northern area of the island, reaching the city of Georgetown and killing 1,680 people. In the period 1971–72 a lava dome was built inside the crater lake, the water of which boiled at 179°F (81.5°C). Soon the dome rose above the level of the water, eventually forming an island with a height of 223 feet (68 m) over the surface of the lake. From April 13 to 26, 1979, the crater was the scene of a series of very intense phreatomagmatic explosions produced by magma coming into contact with the water of the lake. The explosions destroyed much of the lava dome and generated columns of gas and ash that rose up to 12½ miles (20 km) high. The activity later continued with the rebuilding of the lava dome, which forms the island in the crater lake. The precautionary evacuation of the resident population of the island prevented the serious loss of life.

Access and principal attractions. Numerous routes lead out of the city of Georgetown to make their way up the slopes of the volcano, which is an easy hike unless the way is blocked by some sort of activity. The summit of the volcano offers striking views, with the walls of the ancient crater framing a view of the lake in the active crater with the island in the center, composed of a lava dome dating to the eruptions of 1971–72 and 1979.

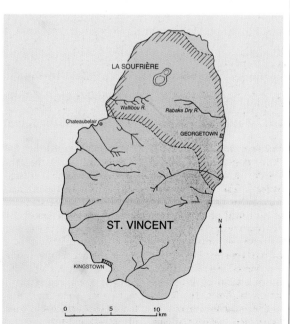

Opposite: One of the columns of ash and vapor produced by the phreatomagmatic eruption of April 1979. The maximum height reached by these columns was about 12½ miles (20 km) over the volcano.
Left: Map showing the area destroyed by the pyroclastic flows caused by the Pelean eruption in 1902 (diagonal lines). Unlike the pyroclastic flows produced only 18 days later on nearby Martinique by Pelée, these pyroclastic flows spread out from the volcano fully 360 degrees, which greatly limited their destructive force. Even so the event caused the deaths of 1,680 people in the city of Georgetown.

LA SOUFRIÈRE

Wallibou R.   Rabaka Dry R.

Chateaubelair

GEORGETOWN

ST. VINCENT

N

KINGSTOWN

0   5   10
km

REFERENCES

GLOBAL VOLCANISM PROGRAM

SMITHSONIAN'S GLOBAL VOLCANISM PROGRAM

The Smithsonian's Global Volcanism Program (GVP), housed in the National Museum of Natural History on the National Mall in Washington D.C., is devoted to better understanding of Earth's active volcanoes and their eruptions. The Smithsonian takes two complementary approaches toward this goal, one a databasing and archival effort that looks back in time over the past 10,000 years, and the other a focus on rapid reporting and dissemination of observations about on-going eruptions.

The GVP's unique retrospective effort towards documenting Earth's eruptive activity began in 1971. Museum volcanologists now count ~1,500 volcanoes that have been active during the past 10,000 years, and have records for more than 8,500 dated eruptions from them. This dataset, which serves as a global resource for scientists, public officials, and others interested in volcanoes and their hazard implications, has been twice published as books titled Volcanoes of the World, and was most recently released electronically as a component of the GVP website.

Complementing the retrospective databasing effort for Earth's active volcanoes and their eruptions, and feeding into it from the modern end, is a focus on reporting of Earth's on-going eruptive activity. The Smithsonian has published global-scale volcanic activity reports since 1968. These reports have been prepared on a monthly cycle since 1975 and have been augmented by weekly reporting (in collaboration with the U S Geological Survey's Volcano Hazards Program) since November 2000. Information flows to the GVP from the Global Volcanism Network, a world-wide group of volcano-watchers, ranging from poorly equipped individual teachers or scientists in developing countries, to fully equipped modern volcanological observatories (listed in the WOVO appendix on pages 328–331) and groups monitoring satellite data for NOAA, NASA, and equivalent agencies in other lands. During the early stages of an eruption anywhere in the world the GVP acts as a clearinghouse of reports, data, and imagery. The monthly reports continue to be published and distributed in hardcopy, but all Smithsonian volcanic activity reports published since 1968 are also available on the GVP website, sorted either by publication date, or chronologically by volcano.

Supporting all of these programmatic efforts is the GVP archive, which houses geologic and topographic maps for most of Earth's active volcanoes, images of volcanoes by Smithsonian and other volcanologists from across the globe, and volcano-related publications, films, and videos. Relevant maps and other archival data are thus on hand for immediate reference during times of volcanic crises.

Smithsonian volcano and eruption data have also been used for scientific and educational applications such as a world map of volcanoes, earthquakes, and plate-tectonic features, CD-ROMs on regional volcanism, and an interactive CD-ROM that sequentially plots volcanic eruptions and earthquakes since 1960 on world and regional physiographic maps, illustrating the dynamic nature of our active planet.

Smithsonian Institution
Global Volcanism Program

P.O. Box 37012
NMNH, MRC-119
Washington, D.C., 20013-7012
USA

Web Site
www.volcano.si.edu/gvp/

E-mail
gvn@volcano.si.edu

GLOSSARY OF VOLCANOLOGICAL TERMS

Ash
Measurement of particle size applied to the finest pyroclastic material, fragments less than 0.08 inches (2 mm) in diameter, blasted into the air by volcanic explosions.

Basalt
Volcanic rock poor in silica but rich in magnesium, iron and calcium; by far the most common volcanic rock.

Base surge
Turbulent low-density cloud of gas with suspended solid debris that expands radially outward from the base of an eruption column. Comparable to the ring that forms at the base of the cloud formed by a nuclear explosion.

Bomb
Fragment of magma varying in diameter from a few inches to several feet ejected into the air during a volcanic explosion; see also Tephra.

Caldera
Large depression produced following an eruption by the collapse of the roof of a magma chamber; usually circular or horseshoe-shaped when viewed from above.

Cinder cone
Cone built up by the accumulation of loose bits of magma (scoria) that fall around a vent or crater after being expelled into the air during moderately explosive activity. If still sufficiently hot they meld when they fall to the ground.

Compound volcano
Volcanic structure composed of two or more vents or lava domes or stratovolcanoes, often formed at different times.

Conduit (volcanic)
Passageway, more or less cylindrical in shape, through which magma rises from the magma chamber to the surface during volcanic activity.

Crater (volcanic)
Bowl-shaped depression or hollow, usually with steep sides, at the summit of a volcano or on its flanks, produced by explosive activity.

Dike
Tabular body of intrusive magma that cuts across layers of a volcanic structure.

Eruption column
Vertical cloud of gas and pyroclastic fragments that forms during explosive eruptions and can reach stratospheric heights.

Fall deposit
Layer of fragmentary volcanic material created by fall from an eruption column or during an explosive eruption.

Fissure volcano
Structure composed of the accumulation of volcanic material on the sides of an eruptive fissure, usually associated with a ridge or rift situation.

Fumarole chimney
Vertical structure with a vent through which volcanic gases are discharged.

Geyser
Intermittent vertical jet of water produced by the heating of

GLOSSARY

Geyser (continued)
underground water by geothermal activity.

Historical time
The time in which events have been recorded by human observers.

Hornito
Small cone or mound produced by the accumulation of spatter ejected by a vent. See spatter cone.

Hot spot
Area on the planet where material from the mantle rises through a plume to reach a moving plate.

Hummocks
Small hills that rise above deposits from a volcanic avalanche.

Hyaloclastite
Fragmentary material of glassy composition produced by the rapid cooling of magma in contact with water.

Ignimbrite
Particular type of pyroclastic-flow deposit containing glassy lenticular structures called flames; ignimbrites can cover thousands of square miles of area.

Lahar
Indonesian word for a volcanic mudflow usually formed when an eruption melts part of a glacier on a volcano; such flows involve large quantities of volcanic material and can move at high speeds.

Lapilli
From the plural for the Latin word lapillus, *meaning "small stone"; small fragments of lava (between 0.08 inches [2 mm]*

and 2½ inches [6.4 cm] across) shot into the air during eruptions when still molten.

Lava
Magma that has reached the surface of the Earth.

Lava dome
Accumulation of lava in the shape of a steep-sided dome over an eruptive vent.

Lava lake
Lake of molten lava inside a crater tens or hundreds of feet in diameter; such lakes can remain active for many years.

Lithic (volcanic)
Fragments of previously formed rocks (volcanic and nonvolcanic) expelled from a volcanic vent during explosive activity and encased in pyroclastic deposits.

Maar
Tuff ring that has filled with water; maars are shallow craters usually with flat bottoms filled with water.

Magma
Molten rock within the earth formed by partial melting of the mantle.

Magma chamber
Reservoir in which magma is accumulated and stored, located within the structure of a volcano a few miles from the surface.

Mudflow
See Lahar.

Neck
Rocky pillar (shaped like a neck) composed of magma that has solidified inside a volcanic conduit and been exposed by the process of erosion.

Nuées ardentes
Generic term, derived from the French for "glowing cloud," applied to pyroclastic flow.

Pillow lavas
Roundish masses of lava produced by effusive underwater eruptions.

Pumice
Frothy volcanic rock, heavily vesicled, formed by the expansion of gas in erupting lava.

Pyroclast
Literally a "fire rock fragment," a piece of material formed by a volcanic explosion or ejected from a volcanic vent.

Pyroclastic flow
Mixture of fragmented volcanic materials and gas traveling at high speeds down the slope of a volcano; formed during an eruption or following the collapse of a lava dome. See also Nuées ardentes.

Pyroclastic-flow deposit
Chaotic deposit of ash, pumice and volcanic rock accumulated in valleys or on slopes and sometimes containing carbonized plant material swept up during the descent. Such deposits can be loose or solid depending on their temperature at the moment of being deposited.

Scoria
Dark, vesicled volcanic rock produced by moderately explosive activity or lava flows and resulting from blobs of gas-charged lava that have cooled in flight.

Shield volcano
Large volcanic structure with long, gentle slopes built up almost entirely from fluid lava flows.

Somma volcano
Type of volcanic structure composed of two volcanic cones, one of which (the more recent) has formed inside the other.

Spatter
Small fragments of lava ejected from a vent; spatter is still molten when it falls to the ground, thus forming cones or hornitos.

Spatter cone
Small, steep-sided mound or cone built up over a vent.

Stratovolcano
Volcano composed of alternating strata, or layers, of material created from lava and pyroclastic flows.

Tephra
Generic term (from a Greek word for "ashes") used for all pyroclastic materials of all sizes when ejected from a volcano.

Tuff
Deposited pyroclastic rock that has consolidated.

Tuff cone
Cone with relatively steep sides built up from the accumulation of fine-grained volcanic fragments produced from explosive activity resulting from the interaction of magma and water.

Tuff ring
Volcanic ring whose height is less than its diameter, built of debris around a volcanic vent located near water and produced by explosive activity resulting from the interaction of magma and water. See Maar.

The World Organization of Volcano Observatories (WOVO), 1996–97

EUROPE

Osservatorio Vesuviano
80056 Ercolano (NA), Italy

Istituto Internazionale di
Vulcanologia – CNR
Piazza Roma, 2
95123 Catania (CT), Italy

Aristototle University of
Thessaloniki, Greece
Faculty of Sciences,
Department of Geology and
Physical Geography
54006 Thessaloniki, Greece

Observatoires Volcanologiques
Institute de Physique du Globe
de Paris
B89 4 Place Jussieu
75252 Paric Cedex 05, France

British Geological Survey
Keyworth
Notts NG12
United Kingdom

Nordic Volcanological Institute
Geoscience Building, University
of Iceland
IS-105 Reykjavik, Iceland

The Sciences Institute
Earthquake Research Branch
Dunhagi 5
IS-107 Reykjavik, Iceland

Departemento de
Geosciecias/Centro de
Vulcanologia
Universidade dos Açores
9500 – Ponta Delgada
Açores, Portugal

Instituto de Meteorologia
Rua C – Aeroporto
1700 Lisbon, Portugal

Central Office for
Observatories in the Azores:
Observatorio Afonso Chaves
Rua Mae de Deus, 58 9500
Ponta Delgada
Azores, Portugal

Departamento de Volcanologia
Museo Nacional de
Ciencias Naturales
Jose Gutierrez Abascal 2
E-28006 Madrid, Spain

Instituto de Astronomia
y Geodesia
Facultad de Ciencias
Matematicas
Ciudad Universitaria
E-28040 Madrid, Spain

Centro Geofisico de Canarias
C/La Marina, Edificio Multiple
E-38001 Santa Cruz de Tenerife,
Spain

Volcanological Station
of the Canary Islands
Spanish Research Council
P.O. Box 195 – 38206 La Laguna
Tenerife, Canary Islands, Spain

AFRICA

Centre de Recherche
en Sciences Naturelles
Lwiro D. S. Bukavo (Sud-Kivu)
Republic of Zaire

Institute de Recherches
Geologiques et Minihres
Antenne de Recherches
Geophysic et Volcanologiques
BP 370 Buea, Cameroon

Observatoire Volcanologique
du Kartala
C.N. de Documentation et de la
R. Scientifique de Comores
BP 169
Moroni, Republique Federale
Islamique des Comores

ASIA AND OCEANIA

Institute of Geological and
Nuclear Sciences Ltd.
Volcanology Programme
Wairakei Research Centre
Private Bag 2000
Taupo, New Zealand

Taranaki Volcano Observatory
Civil Defence Headquarters
45 Robe Street
New Plymouth, New Zealand

Rabaul Volcanological
Observatory
P.O. Box 386
Rabaul, Papua New Guinea

Water and Mineral
Resources Division
Ministry of Energy, Water and
Minerals Resources
P.O. Box G37
Honiara, British Solomon Islands

Department of Geology, Mines
and Water Resources ORSTOM
(French Scientific Research
Institute for Cooperation
Development)
P.O. Box 76, Box 001
Port-Vila, Vanuatu

Volcanological Survey
of Indonesia
Jalan Diponegoro No. 57
Bandung 40122
Indonesia

Philippines Institute of
Volcanology and Seismology
5th and 6th Floors, Hizon Bldg.
29 Quezon Avenue
Quezon City, Philippines

Japan Meteorological Agency
Seismological and Volcanological
Department
1-3-4 Ote-machi, Chiyoda-ku
Tokyo 100, Japan

Earthquake Research Institute
University of Tokyo
Bunkyo-Ku, Tokyo 113, Japan

Hydrographic Department,
Maritime Safety Agency
Ministry of Transport
Tsukiji 5-3-1, Chuo-Ku
Tokyo 104, Japan

Nansei-Toko Observatory for
Earthquakes and Volcanology
Faculty of Science,
Kagoshima University
Korimoto 1-21-35, Kagoshima-shi
890, Japan

Geographical Survey Institute
Ministry of Construction
Kitasato-1, Tsukuba
Ibaraki 305, Japan

Geological Survey of Japan
Environmental Geology
Department
Volcanology Section 1-1-3,
Higashi, Tsukuba
Ibaraki 305, Japan

Sakurajima Volcano
Research Center
Disaster Prevention Research
Institute, Kyoto University
Sakurajima-cho,
Kagoshima-ken 891-14, Japan

Kirishima Volcano Observatory
1489 Suenaga
Ebino-shi
Miyazaki-ken 889-43, Japan

Shimabara Earthquake
and Volcano Observatory
Faculty of Sciences,
Kyushu University
Shimabara-shi
Nagasaki-ken 855, Japan

Research Center for
Seismology and Volcanology
School of Science,
Nagoya University
Furo-cho, Chikusa-ku
Nagoya, 464-01, Japan

Aso Volcanology Laboratory
Kyoto University, Choyo-Mura
Aso-Gun, Kumamoto-ken 869-14
Japan

Asama Volcano Observatory
Nagakura-yama,
Karuizawa-machi
Kitasaku-gun
Nagano Prefecture 389-01, Japan

Kusatsu-Shirane
Volcano Observatory
Tokyo Institute of Technology
Kusatsu, Agatsuma-gun, Gunma
377-17
Japan

Earthquake and
Volcano Observatory
Faculty of Science
Hirosaki University
Bunkyo-cho 3, Hirosaki 036,
Japan

Laboratory for Earthquake
Chemistry
Faculty of Science,
University of Tokyo
7-3-1 Hongo, Bunkyo 113, Japan

Usu Volcano Observatory
Faculty of Science,
Hokkaido University
Sohbetsu-cho, Usu-gun
Hokkaido 052-01, Japan

Institute of Volcanology
FESD USSR Accademie of Sciences
Petropavlovsky
Kamchatsky, 683006, USSR

Institute of Volcanic Geology
and Geochemistry
Far East Division
Russian Academy of Sciences
9, Piip Avenue
Petropavlovsk n Kamchatsky,
683006, USSR

Heilongjiang Wudalianchi
Volcanic Monitoring Observatory
Wudalianchi, Heihe, Heilongjiang
Post Code 164500, China

AMERICAS

Alaska Volcano Observatory
United States Geological Survey
4200 University Drive
Anchorage, AK 99508-4667 USA

University of Alaska
Geophysical Institute
University of Alaska Fairbanks
Fairbanks, AK 99775 USA

Alaska State Division
of Geological and
Geophysical Surveys
794 University Ave., Suite 200
Fairbanks, AK 99709 USA

David A. Johnston
Cascades Volcano Observatory
United States Geological Survey
5400 MacArthur Blvd.
Vancouver, Washington 98661
USA

Long Valley Caldera Monitoring
MS 977
United States Geological Survey
345 Middlefield Road
Menlo Park, CA 94025 USA

The Hawaiian Volcano
Observatory
United States Geological Survey
P.O. Box 51
Hawaii National Park, Hawaii
96718 USA

Centro Universitario
de Investigacio en
Ciencias de la Tierra
Universidad de Colima
Revolucion 132, P.O. Box 380
28000 Colima, Colima, Mexico

Popocatepetl
Volcano Observatory
Centro Nacional de Prevencion
de Desastres
Delfin Madrigal 665, Col.
Pedregal de Santo Domingo
Cayoacan, 04360, Mexico DF,
Mexico

Observatorio Vulcanologico de
Volcan Santiaguito Guatemala
Instituto Nacional de Sismologia,
Vulcanologia, Meteorologia e
Hidrologia
7 Avenida 14-57, Zona 13
Guatemala City, Guatemala

Instituto Nicaraguenza
de Estudios Territoriales
Volcanology Department
Apartado 2110
Managua, Nicaragua

Observatorio Vulcanologico y
Sismologico de Costa Rica
Universidad Nacional
P.O. Box 86-3000
Heredia, Costa Rica

Observatorio Sismologico y
Vulcanologico de Arenal Y
miravalles
Formerly Observatorio
Vulcanologico de Arenal Oficina
de Sismologia y Vulcanologia,
Instituto Costaricense de
Electricidad, Apdo. 10032-1000,
San José, Costa Rica

Observatorio Vulcanologico Y
Sismologico
Instituto de Investigation en
Geociencias Mineraria y Quimica
P.O. Box 1296
Manizales, Colombia

Observatorio Vulcanologico y
Sismologico
Pasto Carrera, 31, 18-07 AA:1795
Pasto, Colombia

Observatorio Vulcanologico y
Sismologico de Popayan
Calle 5B N. 2-14
Popayan (Cauca), Colombia

Instituto Geofisico
Escuela Politecnica Nacional
Casilla 17-01-2759
Quito, Ecuador

Instituto Geofisico del Peru
Arequipa
Urbanizacion La Marina, B12
Cayma, Arequipa, Peru

Southern Andes
Volcano Observatory
Departamento Ciencias Fisicas
Facultad de Ingeneria
Universidad de la Frontera
Campus Andreas Bello, Avda.
Francisco Salazar 01145
Casilla 54-D, Temuco, Chile

Seismic Research Unit
Department of Physics
University of the West Indies
St. Augustine, Trinidad, W.I.

Montserrat Volcano Observatory
c/o Chief Ministeris Office
P.O. Box 292
Montserrat, West Indies

Observatoire Volcanologique
de la Soufrière
Le Houelmont
97113 Gourbeyre
Guadaloupe, France

Observatoire Volcanologique
de la Montagne Pelée
Morne des Cadets
Fondts St. Denis, 07250 Saint
Pierre
Martinique, France

Observatoire Volcanologique du
Piton de la Fournaise
14 RN3 Km 27
97418 La Plaine des Cafres
Reunion, France

330

ANTARCTICA

Mount Erebus Volcano
Observatory
c/o Department of Geoscience
New Mexico Institute of Mining
and Technology
Socorro, New Mexico 87801, USA

Deception Volcanic Observatory
South Shetland Island,
Antarctica
Instituto Antartico Argentinno
Cerrito 1248 Buenos Aires
Museo Nacional de Ciencias
Naturales
Josè Gutierrez Abascal, 2
28006 Madrid, Spain

SCIENTIFIC TEXTS

Abbot, Patrick L.
Natural Disasters
Dubuque: William C. Brown
Publishers, 1996

Blong, Russell J.
Volcanic Hazards
New York: Academic Press, 1984

Bolt, Bruce A., et al.
Geological Hazards
Berlin-New York: Springer-Verlag,
1975

Cas, Ray F., and J. V. Wright
Volcanic Successions
London-Boston: Allen & Unwin,
1987

Catalogue of the Active
Volcanoes of the World,
Including Solfatara Fields
Naples: International Association
of Volcanology and Chemistry of
the Earth's Interior

Fisher, Richard V., and
Grant Heiken
Volcanoes: Crucibles of Change
Princeton, NJ: Princeton
University Press, 1997

Fisher, Richard V., and
Hans-Ulrich Schmincke
Pyroclastic Rocks
Berlin-New York: Springer-Verlag,
1984

Gasparini, Paolo, et al.
Volcanic Seismology
Berlin-New York: Springer-Verlag,
1992

Green, Jack, and Nicholas M.
Short, eds.
Volcanic Landforms and
Surface Features
New York: Springer-Verlag, 1971

Hess, Paul C.
Origins of Igneous Rocks
Cambridge, MA: Harvard
University Press, 1989

Kilburn, Christopher R. J., ed.
Active Lavas
London: UCL Press, 1993

McClelland, Lindsay, et al.
Global Volcanism, 1975–85
Englewood Cliffs, NJ:
Prentice-Hall, 1989

Macdonald, Gordon Andrew,
and Agatin T. Abbott
Volcanoes in the Sea:
The Geology of Hawaii
Honolulu: University of Hawaii
Press, 1970

Ollier, Cliff, and Colin Pain
The Origin of Mountains
London-New York: Routledge,
2000

Prager, Ellen J.
Furious Earth
New York: McGraw-Hill, 2000

Rittman, Alfred
Volcanoes and Their Activity
New York: Interscience
Publishers, 1962

Scarpa, Robert, and R. I. Trilling
Monitoring and Mitigation
of Volcano Hazard
Heidelberg: Springer-Verlag, 1996

Sheridan, Michael F., and
Franco Barberi, eds.
Explosive Volcanism
Amsterdam-New York: Elsevier,
1983

BIBLIOGRAPHY

Siebert, Lee and Simkin, Tom
Volcanoes of the World: an Illustrated Catalog of Holocene Volcanoes and their Eruptions.
Smithsonian Institution, Global Volcanism Program Digital Information Series, GVP-3
(http://www.volcano.si.edu/gvp/world), (2002–)

Simkin, Tom, and Siebert, Lee
Volcanoes of the World, 2nd edition
Tucson, AZ: Geoscience Press, 1994

Venzke, Edward, et al.
Global Volcanism, 1968 to the Present.
Smithsonian Institution, Global Volcanism Program Digital Information Series, GVP-4
(http://www.volcano.si.edu/gvp/reports/), 2002–

Williams, Howel, and Alexander R. McBirney
Volcanology
San Francisco: Freeman, Cooper, 1979

Wohletz, Kenneth, and Grant Heiken
Volcanology and Geothermal Energy
Berkeley: University of California Press, 1992

Wylie, Peter J.
The Way the Earth Works
New York: John Wiley, 1976

POPULAR SCIENCE

Bardarson, Hjalmar R.
Ice and Fire
Hjalmar R. Bardarson, 1991

Brantley, Stephen R.
Volcanoes of the United States
U.S. Geological Survey, 1995

Bullard, Fred M.
Volcanoes of the Earth
Austin: University of Texas Press, 1984

Decker, Robert W., and Barbara B. Decker
Mountains of Fire
Cambridge-New York: Cambridge University Press, 1991

Decker, Robert W., and Barbara B. Decker
Volcanoes
San Francisco: W. H. Freeman, 1989

Francis, Peter
Volcanoes
Harmondsworth: Penguin, 1976

Francis, Peter
Volcanoes: A Planetary Perspective
Oxford: Oxford University Press, 1993

Harris, Stephen
Fire Mountains of the West
Missoula, MT: Mountain Press, 1987

Krafft, Maurice, and Katia Krafft
Volcano
New York: Abrams, 1975

Krafft, Maurice, and Katia Krafft
Volcanoes: Earth's Awakening
Maplewood, NJ: Hammond, 1980

Macdonald, Gordon Andrew
Volcanoes
Englewood Cliffs, NJ: Prentice-Hall, 1972

McGuire, William J., et al., eds.
The Archaeology of Geological Catastrophes
London: Geological Society, 2000

Ollier, Cliff
Volcanoes
Oxford: Basil Blackwell, 1988

Scarth, Alwyn
Volcanoes
College Station, TX: Texas A&M University Press, 1994

Time-Life Books, ed.
Volcano
Alexandria, VA: Time-Life Books, 1982

Van Rose, Susanna
Earthquakes
London: Her Majesty's Stationery Office, 1983

Van Rose, Susanna
Volcanoes
London: Her Majesty's Stationery Office, 1974

REGIONAL GUIDES

Abatino, E.
Vesuvio, un vulcano e la sua storia
Carcavallo, 1989

Allan, Iain, ed.
Guide to Mount Kenya and Kilimanjaro
Nairobi: Mountain Club of Kenya, 1991

Bénard, Roland, and Maurice Krafft
Au Coeur de la Fournaise
Cernay: Editions Nourault/Bénard, 1988

Carrubba, Paolo
A Piedi sull'Etna
ITER, 1992

Chester, David K., et al.
Mount Etna: The Anatomy
of a Volcano
Stanford, CA: Stanford University
Press, 1985

Druitt, Timothy H., et al.
Santorini Volcano
London: Geological Society, 1999

Einarsson, D.
Geology of Iceland
Reykjavik: Mal og menning, 1994

Escritt, T.
Iceland, the Traveller's Guide
Iceland Information Centre, 1990

Franke, Joseph
Costa Rica's National Parks and
Preserves
Seattle, WA: The Mountaineers,
1999

Gasparini, Paolo, ed.
Un Viaggio al Vesuvio
Naples: Liquori, 1991

Giacomelli, L., and R. Scandone
Campi Flegrei Campania Felix
Guida alle escursioni dei
vulcani napoletani
Naples: Liquori, 1992

Gudmundsson, Ari Trausti
Volcanoes in Iceland
Reykjavik: Vaka-Helgafell,
1996

Gudmundsson, Ari Trausti,
and Ragnar Th Sigurdsson
Vatnajokull: Ice on Fire
Reykjavik: Arctic Books, 1996

Institute for the Study and
Monitoring of the Santorini
Volcano
Santorini Guide to the Volcano
I.S. MO.SA.V.

Krafft, Maurice
Guide des Volcans d'Europe
Lausanne-Paris: Delachaux et
Niestlé, 1999

Macdonald, Gordon Andrew,
and Douglass H. Hubbard
Volcanoes of the National
Parks in Hawaii
Honolulu: Hawaii Natural History
Association, 1973
Parc Naturel Régional des
Volcans d'Auvergne
Volcanologie de la Chaine
du Puys
A.R.P.E.G.E., 1983

Porarinsson, Sigurdur
Surtsey, the New Island in the
North Atlantic
Almenna Bokafòlagid, 1964

Prahl, Carlos
Guía de los Volcanes
de Guatemala
Guatemala: Club Andino
Guatemalteco, 1989

Racheli, Gin
L'Isola di Stromboli
Oreste Ragusi Editore

Racheli, G., and O. Ragusi
L'Isola di Vulcano
Oreste Ragusi Editore

Rachowiecki, Rob, and
B. Wagenhauser
Climbing and Hiking in Ecuador
Bradt Publications, 1994

Secor, R. J.
Mexico's Volcanoes
(A Climbing Guide)
Seattle, WA: The Mountaineers,
1993

Spallanzani, L.
Viaggio all'Etna
(Gasparini, P., ed.)
Cuen, 1994

Thorarinsson, S.
Surtsey: The New Island
in the North Atlantic
London: Cassell, 1969

Touring Club Italiano
Parco dell'Etna, guida turistica
Ente Parco dell'Etna, 1993

Williams, David
Iceland, the Visitor's Guide
London: Stacey International,
1985

PHOTO CREDITS

Alessandro Agostini: 55, 252b
James Allan: 244, 247
Norm Banks: 204, 272, 275, 296, 308
Franco Barberi: 20–21
Costanza Bonadonna: 311
Philippe Bourseiller: 60 sx
Steve Brantley: 226
Steve Carey: 250, 251
Tom Casadevall: 179, 180, 181
Mauro Coltelli: 30, 34, 37a, 37b, 44b, 95, 110, 112, 114, 117
Jean Paul Degas: 126, 128a, 128b, 139
Paolo Fiorini: 285
Richard Fiske: 240, 314, 316
Carl Fries: 246
Rick Hoblitt: 248a
Bruce Houghton: 53, 65b, 168
Warren Huff: 245
Krafft/HOA-QUI/Franca Speranza: 13, 15, 16, 22, 26, 33, 44a, 47, 50c, 50b, 56, 57, 64, 72–73, 77, 90–91, 130, 132, 134, 136, 140, 154, 156, 164, 166, 172, 174, 177, 178, 203, 212, 214, 273, 303
Nicolas Lahuer: 170
Luca Lupi/Vulcano Esplorazioni: 24a (1, 2, 3), 25b, 37c, 42, 43c, 50a, 51a, 51b, 58c (1, 2, 3), 59b, 63a, 65c (1, 2, 3), 86, 87, 89, 92a, 94a, 96a, 97b, 101, 102, 104, 105, 108, 112, 132, 133, 135, 141, 150, 152, 182b, 265, 281, 291
José Manuel Navarro: 60c, 46c
Dan Miller: 206, 207a, 207b, 208, 222, 223, 225, 232, 235, 248b 277
Sabrina Mugnos: 14, 23b, 153
Nagasaky Photo Service: 199
Chris Newhall: 188, 189, 190, 193b, 260
Michael Ort: 66
Paolo Papale: 23a, 89, 96c, 97, 278
Don Peterson: 176, 242
Tom Pierson: 16, 202
Sergio Ramazzotti: 262a
William Rose: 55, 203b, 221, 252a, 253, 254, 255
Mauro Rosi: 24b (1, 2, 3), 25b, 45b, 60c, 61a, 61b, 62, 63c, 66, 83, 93a, 93c, 100, 104, 110, 111, 120, 151, 157, 163, 192, 193c, 195, 216, 218, 219, 246, 274, 276a, 280, 286, 288
Guido Alberto Rossi/Image Bank: 106, 109, 118
Lee Siebert: 203, 220, 224, 230, 236, 238, 256
Bruce Spainhower: 226
Marco Stoppato: 8–9, 11, 25a, 31, 32a, 32b, 43a, 43b, 44c, 45a, 45c, 46a, 58a, 59a, 59c, 67a, 67c, 67b, 68, 69, 70 (1, 2, 3, 4), 71c, 71b (1, 2, 3), 74, 76, 78, 79, 113, 124, 130, 131, 138, 141, 144, 145, 146, 158, 159, 264, 266, 267, 268, 269, 270, 271, 292, 294, 302, 304, 306a, 306b, 307, 312
Shinji Takarada: 198
Alessandro Tibaldi: 74, 109
Lyn Topinka: 234
Hugo Torres: 276b, 282, 284, 290
Tadahide Ui: 80–81, 196, 200
Jim Vallance: 38–39, 228
Renato Valterza: 298
Luigi Vigliotti: 35, 92b, 94b, 162a, 162b, 182a, 184, 185
G. José Viramonte: 258, 259, 261, 262b, 300, 301
Giorgio Virgili: 85, 94b
Rick Wunderman: 122, 124

MAPS, DRAWINGS, DIAGRAMS

The topographical maps and geological cross-sections that appear throughout this book with identical background colors come from a variety of sources and were in many cases supplied by volcanologists throughout the world who contributed to this book. They come from scientific magazines, universities, observatories, bulletins, national parks, mountain-climbing books, etc.

All the other drawings, diagrams and maps used in the book, particularly those used with the volcano entries themselves, including the icons used for symbols (geodynamic setting, volcanic structure), were made by Luca Lupi (Vulcano Esplorazioni) and then reworked on computer by Studio L'Atelier of Modena. The only exceptions are the drawings and maps on pages 119, 121, 183, 209, 213, 279, 283, 287, 289, 293 and 295.

ACKNOWLEDGMENTS

The authors would like to give particular thanks to the following for their contributions:

Mauro Coltelli (researcher with the International Institute of Volcanology of the CNR of Catania) for assistance in the preparation of the material on volcano monitoring and for providing images of the volcanoes of Sicily and Hawaii.

Mauro Di Paola for reviewing the introductory texts and making many useful observations, and for making available his large collection of specimens of volcanic activity, some of them extremely rare and from every part of the world.

Lee Siebert (Smithsonian Institution, Washington) for furnishing a large number of images from the Global Volcanism Program archive of world volcanic activity.

Gianni Macedonio (professor of geophysics at the Vesuvius Observatory) and Augusto Neri (researcher with CNR in Pisa) for providing the series of images from a computer simulation of a volcanic eruption (page 97) and the computer-generated images of the topography of Etna (page 115).

Joan Marti (Institute of Earth Sciences, CSIC, Barcelona) for providing the DEM computerized image of the island of Tenerife (page 129).

Jim Luhr, Director of the Global Volcanism Program, for furnishing information on the GVP.

335

DATE DUE

4-16-08			

HIGHSMITH #45115

X 1/4/11

DATE DUE

10-8-08			
5-6-09			
11-28-11			
10-25-12			
5/13			
1/25/16			

HIGHSMITH #45115